SpringerBriefs in Biology

For further volumes:
http://www.springer.com/series/10121

Michael A. Huffman • Naofumi Nakagawa
Yasuhiro Go • Hiroo Imai • Masaki Tomonaga

Monkeys, Apes, and Humans

Primatology in Japan

 Springer

Michael A. Huffman
Primate Research Institute
Kyoto University
41-2 Kanrin, Inuyama
Aichi 484-8506
Japan

Yasuhiro Go
Primate Research Institute
Kyoto University
41-2 Kanrin, Inuyama
Aichi 484-8506
Japan

Masaki Tomonaga
Primate Research Institute
Kyoto University
41-2 Kanrin, Inuyama
Aichi 484-8506
Japan

Naofumi Nakagawa
Laboratory of Human Evolution Studies
Graduate School of Science
Kyoto University
Kitashirakawa-Oiwakecho, Sakyo-ku
Kyoto 606-8502
Japan

Hiroo Imai
Primate Research Institute
Kyoto University
41-2 Kanrin, Inuyama
Aichi 484-8506
Japan

ISSN 2192-2179   ISSN 2192-2187 (electronic)
ISBN 978-4-431-54152-3   ISBN 978-4-431-54153-0 (eBook)
DOI 10.1007/978-4-431-54153-0
Springer Tokyo Heidelberg New York Dordrecht London

Library of Congress Control Number: 2012948989

Printed on acid-free paper

Springer is part of Springer Science+Business Media (www.springer.com)

Foreword

While the twentieth century was the century when researchers tried to discover "the general basic principles of organisms", the twenty-first century is expected to be the century when researchers try to understand "the evolution and diversity of organisms" on the basis of such general principles of organisms by integrating various disciplines such as morphology, physiology, and ecology.

The chief difficulty in studying "the evolution and diversity of organisms" lies in the fact that we have to consider factors at various levels ranging from the genome to the ecosystem. As taking various factors into account may cause a loss of focus, traditional studies have been restricted to analyzing only one individual level or factor. However, unfortunately, the current research and education system based on such a compartmentalized approach is inadequate for incisively studying "the evolution and diversity of organisms."

In order to solve these problems, we should strongly emphasize the necessity for joint studies and integration of the education programs between micro-level biology (genomic science, evolutionary developmental biology, genetic science, cell biology, neurobiology, molecular physiology, molecular evolutionary study) and macro-level biology (primatology, anthropology, ethology, environmental biology, evolutionary taxonomy, and so on) to young biologists. We launched a new education program in Kyoto University, called "Global COE program for Evolution and Biodiversity Research", to promote such integrative studies at the various levels, and have succeeded in initiating novel currents of study of biodiversity that led rather than followed those in the rest of the world. To this aim, we decided to publish six books in "SpringerBriefs in Biology" which we hope will stimulate interest among young biologists about such novel approaches to the study of evolution and diversity of organisms in the world.

In this book, we introduce primate research ranging from field studies to personal genome sequencing. In Japan, primatology was initiated by the group of Dr. Kinji Imanishi, an Emeritus Professor of Kyoto University, in late 1940s. Firstly, they started to observe wild Japanese macaques in Koshima Islet, southern Japan, and elucidated their social structure and life history by distinguishing and observing each individual. Their discovery of the "cultural behaviors" of Japanese macaques and their original method (distinguishing each individual) made a big impact on primatologists around the world. Next, they sent expedition teams to Africa to observe wild great apes, such as chimpanzees and gorillas, in the 1960s. They observed chimpanzees in the wild and succeeded in provisioning wild chimpanzees in the Mahale Mountains, Tanzania, for the first time. Laboratory research on chimpanzee behavior in Japan was started mainly by psychologists in the Primate Research Institute (PRI), Kyoto University, which was founded in 1967 in Inuyama, central Japan. Recently, we have incorporated personal genome analysis into both fieldwork- and laboratory-based research. We hope readers will enjoy the new trends of primate research described here, and their implications regarding the nature of humans.

Kiyokazu Agata
Professor, Department of Biophysics, Kyoto University
Project Leader of Kyoto University Global COE program
"Evolution and Biodiversity"

Preface

The goal of this book is to introduce to the reader unfamiliar with primatology in Japan, examples of some of the representative work carried out by scientists at the country's premier institution for primate studies, Kyoto University, and to provide three unique multidisciplinary approaches to understanding the age old question, where did we come from and what makes us unique or similar to our primate ancestors? Kyoto University is one of the world's oldest and arguably most active centers of primatological research today. Research on primates in Japan began in 1948 with the pioneering work on the country's own endemic species the Japanese macaque by researchers based in Kyoto University's Faculty of Science. Soon thereafter, scientists expanded their research interests to Africa, Asia and South America. Many of the standard practices of fieldwork, such as individual identification and long-term studies are but a few of the early contributions made by primatologists in Japan to primatology. The early studies contributed much to our current understanding of primate social organization, social structure and the role of social behavior on inter-individual relationships and mating systems.

Laboratory based research has also long been a strength of Kyoto University and over the years, increasing sophistication in the areas of molecular biology and comparative cognitive psychology have been brought to the field of primatology. A world-renowned example is that of the Ai Project and more recently genomic studies undertaken here at the Primate Research Institute.

The first chapter by Naofumi Nakagawa focuses on the cultural diversity of social behavior in the Japanese macaque. The first example of cultural behavior in animals, "potato washing", comes from pioneering work on Japanese macaques. In the first chapter of this volume, research on primate culture, in particular the work on Japanese macaques is summarized, then a new example, arguably the first example of a culturally transmitted social convention called "hug-hug" is presented.

The second chapter by Michael A. Huffman introduces our current knowledge of self-medication in primates, based largely, but not limited to, long-term study of wild chimpanzees at Kyoto University's longest chimpanzee field in Africa, Mahale, in Tanzania. The suite of behavioral adaptations to parasite infections documented first in chimpanzees is compared with our current knowledge of self-medication in other great apes species, other primates and other non-primate animal species for which evidence is currently available.

The third and final chapter by Yasuhiro Go, Hiroo Imai and Masaki Tomonaga describes the ambitious efforts to combine cognitive science and genomics into an exciting new discipline called "Comparative cognitive genomics". They provide an overview of recent advancements in chimpanzee comparative cognition, the construction of a chimpanzee genomic database and comparative genomic studies at the individual level looking into factors affecting personality and individuality.

It is our privilege and great pleasure to present to you some recent and long-term representative studies focusing on aspects of behavioral and molecular evolution in primates that offer a view into the evolution of our species. We hope you will enjoy reading about this work and that it will stimulate the reader to learn more about primatology and the many, varied contributions made by Kyoto University over the years.

Inuyama, Japan   *Michael A. Huffman*

Contents

Chapter 1
Cultural Diversity of Social Behaviors in Japanese Macaques

Abstract We humans have a variety of cultures that cannot be categorized strictly as material culture. In primatology, however, these cultures of social behavior or "social cultures" that represent particular styles of sociality permeating an array of social behaviors, have been largely ignored because researchers have sought to shed more light on material cultures instead. However, recently inter-population differences in behavioral patterns of an embracing behavior (called "hug-hug") that functions to reduce tension have been found in wild Japanese macaques. In one population, the macaques rock their bodies back and forth in a ventro-ventral position, whereas in another population the embracing occurs in one of three positions—ventro-ventral, ventro-lateral, or ventro-dorsal—and is accompanied by kneading the fur by opening and closing their palms. On the other hand, "hug-hug" has never been observed in some populations or groups, irrespective of long-term intensive observations. Moreover, a less despotic, more tolerant social relationship among females, being characterized by a much larger resting cluster formation, less frequent aggression accompanied by biting, and more frequent counter-aggression than typically reported for Japanese macaques has been rediscovered in another population. All these inter-population differences were considered to be cultural differences. Lastly, behavior-related candidate genes as alternative explanation of the tolerant society will be discussed.

Keywords Cultures • Embracing behavior • Inter-population differences • Japanese macaques • Social behavior • Social structure

1.1 Cultural Behaviors That Have Shed Light on Non-human Primates

What kinds of behavior do you associate with animal culture? Many Japanese would take the example of sweet-potato washing in Japanese macaques, owing to its introduction in primary school Japanese language (*kokugo*) textbooks. In 1952, soon after

Japanese primatological research was begun by a research team from Kyoto University (composed of Masao Kawai, Shunzo Kawamura, and Jun'ichiro Itani, under the guidance of Kinji Imanishi), the story was born on Koshima Islet, Miyazaki Prefecture, in southern Japan (Kawai 1965). In the midst of habituating the macaques to human presence by providing them with sweet potatoes and wheat in order to be able to observe them closely, a 2-year-old female named *Imo* ("potatoe" in Japanese) started to wash the sand off pieces of sweet potatoes with the fresh water of a stream. She also learned to add a salty taste by washing them in seawater. Consequently, this behavior was transmitted to her playmates, mother, and other affiliates, and eventually spread to the rest of the group. At present, after more than 60 years later, most of the current group members perform *Imo*'s sweet-potato washing technique. New behaviors were invented, socially transmitted to other individuals, propagated to the rest of the group, and finally transmitted across generations. Thus, it is correct to say that they have cultural behavior.

Others might be reminded of the inter-population differences of tool-use behavior in wild chimpanzees. For chimpanzees in Mahale, Tanzania, East Africa, where long-term research has been conducted since successful provisioning in 1966 by Toshisada Nishida, one well-known tool-use behavior is ant fishing (Nishida 1973; for updated information, see Nishie 2011). The chimpanzees bend a nearby twig, strip it of leaves, and insert it into the opening of an ant nest built into a tree trunk or large branch. Next they pull the stick tool back out and eat the worker ants that bite and cling to it. On the other hand, ant fishing was first observed in 2003 in Bossou, Guinea, West Africa (Yamamoto et al. 2008), where long-term research has also been conducted since 1975 by Yukimaru Sugiyama. In Bossou, a customary tool-use behavior is nut cracking (Sugiyama and Koman 1979). The chimpanzees puts a seed from the oil palm (*Elaeis guineensis*) onto a large flat stone that functions as an anvil, and cracks it open using a small hammer-like stone, eating the edible kernel inside the hard shell. In Taï, Ivory Coast, instead of oil palm seeds, the nuts of the coula (*Coula edulis*) and panda (*Panda oleosa*) trees are cracked using a stone or a hammer-like log on a stone or root (Boesch and Boesch 1984). Surprisingly, no nut cracking has been observed in either Mahale or Gombe, Tanzania, despite the abundance of oil palm trees and stones. Even in this case, when the environmental conditions for the occurrence of the behavior are the same, some behaviors can only be observed in one site, and not the other. This evidence lends support to the theory of culture in chimpanzees, that is, a behavior was invented, and then transmitted socially to other individuals.

In 1999, a landmark paper entitled "Cultures in chimpanzees," by Andrew Whiten, a cognitive scientist from the University of St. Andrews in the UK, and the leaders of long-term research sites on wild chimpanzees, including Nishida and Sugiyama, was published in one of the top scientific journals, "*Nature*" (Whiten et al. 1999). Sixteen out of thirty-nine behaviors described as being cultural in this paper were tool-use behaviors used for acquiring foods, such as ant fishing, nut cracking, ant dipping, termite-mound digging, and algae scooping. Moreover, by including behaviors that used object-manipulation, such as courtship or aggressive displays using leaves or stones, and leaf grooming (squashing ectoparasites captured during grooming on leaf), 37 different behaviors were listed.

1.2 Forgotten Cultures of Social Behaviors

Let us turn to human cultures. We have a variety of cultures: not all of them can be categorized as material culture. Some cultural behaviors are primarily unrelated to food and/or object-manipulation, that is, religion, social customs, and institutions. Language itself is a product of culture, in the sense that language plays an important role in creating culture by mediating social transmission. Language, especially spoken language, is used in a social context. Let us take for example a greeting that is exchanged in an encounter between two people. Even at similar times of the day, these salutations are different not only between countries but also locally and in different locations within a country. The gestures accompanying those words also vary. Although hand shaking from Western society has become entrenched in Japanese greetings, this was traditionally done by bowing. Japanese generally hesitate to hug each other or exchange kisses on the cheek. On the other hand, Westerners would have trouble kissing feet, as done in some Asian countries, unless it was to greet the Pope. In contrast, today some Japanese find it strange when a Westerner bows to them, holding both hands together in front of their chests. Many westerners do not understand that this gesture was transmitted from India along with Buddhism to some parts of Southeast Asia, such as Thailand, but not to East Asia and Japan.

Recently, a few examples of cultural differences in social behavior have been observed in non-human primates. Michio Nakamura, Kyoto University, found it strange that studies on culture in chimpanzees are biased toward material culture; therefore, he decided to focus on the cultural differences in social behavior of wild chimpanzees. He focused on two behaviors specifically: the grooming hand-clasp and the social scratch. The grooming hand-clasp is a behavior where two partners grasp each other's hands—either their right or left—and raise these mutually clasped arms above their heads while mutually grooming the underarm of their partner with the free hand. This behavior is observed frequently in Mahale but has never been observed in Gombe, just 170 km to the north. It is also customary in Kibale and Kalinzu, Uganda but has never been observed in Budongo, Uganda. In West Africa, it is infrequently observed in Taï but never in Bossou (Nakamura 2002; Nakamura and Uehara 2004; Nakamura and Nishida 2006). Social scratch behavior is simply the scratching of a partner's back. It occurs in Mahale and Kibale, but has never been observed in Gombe (Nakamura et al. 2000). More interestingly, the social scratch behavior has subtle differences depending on the area. Chimpanzees in Kibale scratch using their fingers to "poke" the body of their partners, whereas at Mahale they use their flexed fingers to "stroke" the body (Nishida et al. 2004).

Other well-known examples of cultural differences in the social behavior of non-human primates are those observed in white-faced capuchin monkeys (*Cebus capuchinus*) (Perry et al. 2003). "Hand-sniffing" is a behavior consisting of one monkey covering its own face with its partner's hand or foot, and inhaling repeatedly for over a minute with its eyes closed. Another, the "sucking of body parts" consists of one monkey sucking its partner's fingers, heels, ears, and tails for up to an hour or more. The "finger-in-mouth game" is a somewhat dangerous social game, where one monkey inserts its fingers into the partner's mouth. The partner then bites down hard enough that the finger cannot be easily removed, while the owner of the finger struggles to pry their partner's mouth open to remove their finger safely. These behaviors have been known to vary in frequency among different groups and over time in four study sites in Costa Rica.

However, what about the cultural differences in social behavior of Japanese macaques, made famous for its cultural behavior during the early stages of Japanese primatology? Inter-population differences in social behavior such as paternal behavior and mounting behavior were investigated in terms of their cultural differences across sites at that time (Kawamura 1965), but has since never been rigorously tested. On the other hand, "stone-handling," a kind of solitary play, has been intensively and extensively examined to provide evidence for culture in primates, and was first recorded in 1979 in a 3-year-old female (named *Glance*-6476) in the Arashiyama E group, Kyoto Prefecture, by Michael A. Huffman from Kyoto University (Huffman 1984; see Huffman et al. 2010 for an up-dated review).

1.3 Culture for Embracing Behaviors in Japanese Macaques: Prologue

In October 1984, I began conducting field research on the feeding ecology of Japanese macaques that inhabited Kinkazan Island, Miyagi Prefecture, in northern Japan, as a Master's course student of the Primate Research Institute (PRI) of Kyoto University (see Nakagawa 1989, 1990, 1999). One day, I noticed a strange behavior. An adult female approached another, and sat down next to her, and then proceeded to embrace the her ventro-ventrally with both arms in a sitting position, and then they started to rhythmically rock their bodies back and forth. Both of them pushed their lips out in a rhythmic open-close movement (i.e., lip smacking) and emitted the "girney" vocalization ("*ngu-ngu-ugu*", Fig. 1.1). I had never observed or even heard of such a behavior in wild Japanese macaques at Arashiyama, Takasakiyama (Oita Pref.), Koshima, or Yakushima (Kagoshima Pref.), although my observation period had been too short to confirm its absence at that time. I decided to accumulate the basic data on this behavior, such as participants involved, and the behaviors that occurred before and after, although this behavior occurred very infrequently. In October 1997, after 13 years, Yukiko Shimooka, my junior colleague from PRI, began to intensively investigate the possible functions of this behavior. She named it "hug-hug," and concluded that "hug-hug" functions to reduce tension between participants and facilitates smooth exchange in allo-grooming in her Master's thesis for Kyoto University. In September 2005, that is, after a period of 8 years, I had just started conducting field research on the mating behavior of Japanese macaques in Yakushima as a cooperative project with many primatologists, including Shimooka, Miki Matsubara, another junior colleague of mine from PRI, and my student Mari Nishikawa. It was then that I happened to observe the Yakushima macaques displaying a version of "hug-hug" that I had never heard of or observed before. Its pattern

Fig. 1.1 "Hug-hug" in Kinkazan: one macaque embraces another ventro-ventrally with both arms in a sitting position, and rocks their bodies back and forth

seemed to be slightly different from that of the Kinkazan macaques. Therefore, we decided to collect data comparable to Shimooka's, despite some time restrictions owing to the nature of this endeavor as a sub-project.

1.4   Inter-population Differences in Behavioral Patterns and Presence or Absence of "Hug-Hug"

I compared the behaviors that immediately occurred before "hug-hug" in Kinkazan and Yakushima. Consequently, intermissions of allo-grooming and antagonistic behaviors accounted for approximately 50% of the behaviors, whereas a simple approach not preceded by these behaviors was seen for the remaining cases at both sites. There was no significant difference in the behaviors just before "hug-hug" between the two sites. Almost all the "hug-hug" behaviors were followed by allo-grooming at both sites, and there was no significant difference in the behaviors displayed just after "hug-hug" between the two sites. Thus, we concluded that "hug-hug" in Yakushima was a homologous behavior to that in Kinkazan, which functioned to reduce tension (Nakagawa et al. in preparation).

Despite functional similarities, there were structural differences in "hug-hug" between the two sites. The first difference is in the embracing position. In addition to a ventro-ventral position like in Kinkazan, Yakushima macaques embraced in the ventro-lateral position as well. Embracing in a ventro-dorsal position was also observed once. A second difference is that Yakushima macaques would knead each other's fur by opening and closing their palms during the course of "hug-hug" instead of rigorously rocking their bodies as observed in Kinkazan (Nakagawa et al. in preparation) (Fig. 1.2).

Fig. 1.2 "Hug-hug" in Yakushima: one macaque embraces in a ventro-lateral position and kneads the other's fur by opening and closing its palm (still frame from the videotape taken by Mari Nishikawa)

In November 2008, I happened to observe "hug-hug" with a mixture of these two behavioral patterns in the M87 group of Shimokita Peninsula, Aomori Prefecture where Haruka Taniguchi, my student, was conducting field research. Shimokita macaques exhibited "hug-hug" in three types of positions (Yakushima type), but they were accompanied by body-rocking instead of fur-kneading (Kinkazan type).

Figure 1.3 shows the populations of the Japanese macaques where "hug-hug" has or has not been observed. Hakusan, Ishikawa Prefecture, has been added to those in which "hug-hug" has been observed, though the behavioral pattern is not yet described. On the other hand, as a result of a questionnaire survey targeting long-term field researchers, "hug-hug" has never been observed in three provisioned populations, Arashiyama, Takasakiyama, and Katsuyama (Okayama Pref.) (Nakagawa et al. 2011). Although "hug-hug" appears to occur only in wild non-provisioned populations, and not in provisioned ones, this is not the case. "Hug-hug" has been observed in four out of six groups in the Kinkazan population, though never in the remaining two groups. Information for one (C2 group) of the two groups is reliable, owing to the 7-year observation period of my student, Tatsuro Kawazoe.

There is no doubt that the inter-population differences in the presence or absence of "hug-hug" can be explained neither by environmental nor genetic factors. One cannot imagine that local differences in a behavioral pattern like "hug-hug," affecting minute qualities, such as the position of embracing or the presence of the open-close movement of the palms, can be attributed to environmental or genetic factors. Therefore, the most plausible explanation for these so far is that they are cultural differences. This behavior is usually exhibited by adult females, though sometimes also by juveniles or infants, including males. It is not until social learning of this behavior can be proven that "hug-hug" can be established as being a cultural social behavior in a strict sense.

Fig. 1.3 Locations of wild non-provisioned populations of the Japanese macaques where "hug-hug" has been observed (*solid triangles*), provisioned ones where "hug-hug" has not observed (*blank circles*), and other provisioned ones referred to this article (*cross*)

1.5 "Social Cultures" Among Non-human Primates

Sapolsky (2006) defined "social cultures" as a particular style of sociality permeating an array of behaviors, creating an assemblage of traits that fulfills the criteria for culture. Japan has been considered by some to be a representative of collectivism culture, wherein people are mainly motivated by the norms and duties imposed upon them by the collective entity. On the other hand, Western countries tend to have cultures of individualism, wherein people are motivated by their own preferences, needs, and rights, giving priority to personal rather than group goals (Triandis 1995). The Japanese and overseas media have praised the Japanese for supporting each other and keeping social order without committing crimes, such as looting, even under critical conditions like The Great East Japan Earthquake of 11 March 2011. For example, a French historian and demographer, Emanuel Todt, reportedly said that this collectivism culture in Japan is a driving force for revival (Todt and Mikami 2012).

The best example of social culture in wild non-human primates involves a population of anubis baboons (*Papio hamadryas anubis*) in the Masai Mara Reserve, Kenya (Sapolsky and Share 2004; see also Sapolsky 2006 for review). The forest group exhibited a textbook example of despotic society—aggressive, with a linear dominance hierarchy, and male-dominated—at the onset of a long-term study in 1978.

During the mid-1980s, approximately half the group males, especially more aggressive males, died from bovine tuberculosis and atypically less aggressive males survived. As baboon males leave their natal group just before puberty, by the mid-1990s, none of the males who had resided in the group a decade before remained. Thus, a social structure of less aggression over subordinate males, and higher affiliation with females was adopted, even by new males who joined the group. In addition, indices of chronic stress, such as resting levels of cortisol and the frequency of anxiety-related activity, were not observed among the subordinate males. This social style has been transmitted across generations for nearly two decades.

At the early stages of research, Japanese primatologists focused on the intraspecific differences in the social structure of Japanese macaques. Kawai (1964) employed the following six social indices to categorize the social structure of Japanese macaques: (1) tolerance of leader males (i.e., high-ranking males); (2) social tension; (3) strictness of dominance hierarchy among males; (4) spatial position of sub-leader males (i.e., middle-ranking males); (5) gregariousness of the group; and (6) the degree of double-layered (central and peripheral) spatial structure within a group. By summing up the scores in each social index (total score was called ITS), he categorized 18 groups of provisioned Japanese macaques into three types: Type J ($2 < ITS <= 6$), Type G ($-2 <= ITS <= 2$), and Type A ($-6 <= ITS < -2$). The Takasakiyama group is representative of Type J, which was characterized by intolerant leader males, high social tension, a strict dominance hierarchy, peripheral location of sub-leader males, low gregariousness of the group, and a clear double-layered spatial structure. In contrast, the Kankakei group at Shodoshima Island (Ehime Pref.) is representative of Type A, which was characterized by tolerant leader males, low social tension, loose dominance hierarchy, central location of sub-leader males, high gregariousness, and an unclear double-layered spatial structure. One likely explanation for these differences in social structure was cultural differences—the generation brought up under the influence of high-ranking or elder individuals with a certain personality type would have socially transmitted their personality.

Recently, Zhang and Watanabe (2007) uncovered a forgotten treasure from early Japanese primatologists. They focused on the extra-large resting cluster formation of the Chosikei SA and SB groups on Shodoshima Island. In winter, the mean cluster size in the SA and SB groups were 17.1 and 15.9, respectively, which were significantly larger than those in Takasakiyama B (4.5) and C groups (4.7), although there was no consistent difference in group size between the two populations. The cluster size in Shodoshima became larger under lower temperatures. In contrast, those of Takasakiyama did not increase owing to temperature change, despite no significant differences in mean and minimum temperature between these two study sites. The percentage of adult females was 43% in extra-large clusters comprising less than 51 individuals in Shodoshima. Surprisingly, a maximum of 137 individuals have been seen to huddle in one cluster (Fig. 1.4). More interestingly, adult females in Shodoshima displayed intense aggression accompanied infrequently by biting, but displayed counter-aggression more frequently than did females at Takasakiyama. It follows from the above discussion that Japanese macaques on Shodoshima Island have a less despotic, more tolerant social relationship among

Fig. 1.4 The Chosikei SB group at Shodoshima Island forms an extra-large resting cluster

females than is typically reported for this species. This tolerant trait has been continuously seen for at least half a century in Shodoshima. However, those that were relocated to other habitats did not form very large clusters according to Zhang (personal observation) and Hayashi (personal communication in Zhang and Watanabe 2007). In addition, it is difficult to explain the extra large clusters as an adaptive behavior against cold, because Shodoshima is relatively warm among the range of habitats for Japanese macaques. For the above reasons, Zhang and Watanabe (2007) concluded that such a long-lasting tendency of tolerant social relationships was evidence for social culture among Japanese macaques in Shodoshima.

Tolerant social relationships were also found in the Awajishima Island group (Hyogo Pref.). Koyama et al. (1981) examined the difference in social tension as well as gregariousness in seven provisioned groups. The degree of gregariousness was measured by the number of individuals, which were in a circle of 8 m in diameter, where a fixed number of artificial foods were evenly scattered. The level of tension was measured by the frequency of agonistic vocalizations emitted by individuals in the circle. Overall, a significant positive correlation between the degree of gregariousness and the level of tension was found. The Awajishima Island group, however, was the exception: a low level of tension relative to gregariousness was found. Nakagawa (2010) depicted Yakushima macaques as having tolerant social relationships, because several of their social traits are consistent with those of tolerant macaque species, such as Tonkean (*Macaca tonkeana*), stump-tailed (*M. arctoides*), and Barbary macaques (*M. sylvanus*): high conciliatory tendency, low rate of coalition by kin, no "youngest ascendancy," and no rank-related difference in birth rate and grooming directions (cf. Thierry 2000). However, instead of cultural differences, differences in behavior-related candidate genes may explain such a social difference. Genes related to low aggressiveness (short allele in monoamine oxidase A and long allele in androgen receptor) were found more frequently in the tolerant Awajishima populations than another seven populations, including Takasakiyama, Koshima and Arashiyama (Inoue-Murayama et al. 2010; see also Chap. 3), although genetic data from Yakushima and Shodoshima have not yet been analyzed. Further research is needed to determine whether social differences in Japanese macaques have such a genetic component (see Nakagawa 2010 for details of the intra-specific differences in the social structure of Japanese macaques and alternative explanations for it).

Recently, evolutionary biologists and anthropologists do not think that culture and genes are mutually exclusive, but may have possibly co-evolved. Analysis of data from the human genome have revealed a great deal of genes that have experienced recent positive selection, many of which show functions that suggest that they are responses to human cultural practices (Laland et al. 2010). One of the most famous examples is pastoralism, which culturally spread prior to lactase persistence, the genetic trait conferring the ability to digest the milk sugar lactose in adults. This trait is likely to have conferred a selective advantage in some pastoralist societies (Burger et al. 2007; Tishkoff et al. 2007). Even in societies of non-human primates, such gene-culture co-evolution might have occurred. For example, low aggressive genes might have been selected since tolerant societies form culturally. A cooperative project between traditional long-term individual-based field research and modern personal genomics will enable us to prove such a fancy scenario in the future.

References

Boesch, C., & Boesch, H. (1984). Possible causes of sex differences in the use of natural hammers by wild chimpanzees. *Journal of Human Evolution, 13*, 415–440.

Burger, J., Kirchner, M., Bramanti, B., Haak, W., & Thomas, M. G. (2007). Absence of the lactase-persistence-associated allele in early Neolithic Europeans. *Proceedings of the National Academy of Sciences of the United States of America, 104*, 3736–3741.

Huffman, M. A. (1984). Stone-play of *Macaca fuscata* in Arashiyama B troop: Transmission of a non-adaptive behavior. *Journal of Human Evolution, 13*, 725–735.

Huffman, M. A., Leca, J.-B., & Nahallage, C. A. D. (2010). Cultured Japanese macaques: Multidisciplinary approach to stone handling behavior and its implications for the evolution of behavioral tradition in nonhuman primates. In N. Nakagawa, M. Nakamichi, & H. Sugiura (Eds.), *The Japanese macaques* (pp. 191–219). Tokyo: Springer.

Inoue-Murayama, M., Inoue, E., Watanabe, K., Takenaka, A., & Murayama, Y. (2010). Behavior-related candidate genes in Japanese macaques. In N. Nakagawa, M. Nakamichi, & H. Sugiura (Eds.), *The Japanese macaques* (pp. 293–301). Tokyo: Springer.

Kawai, M. (1964). *Ecology of Japanese macaques [Nihonzaru no Seitai]*. Tokyo: Kawadeshobo [in Japanese].

Kawai, M. (1965). Newly acquired pre-cultural behavior of the natural troop of Japanese monkeys on Koshima Islet. *Primates, 6*, 1–30.

Kawamura, S. (1965). Sub-culture in Japanese macaques. In S. Kawamura & J. Itani (Eds.), *Monkey: Sociological studies* (pp. 237–289). Tokyo: Chuoukouronsha [in Japanese].

Koyama, T., Fujii, H., & Yonekawa, F. (1981). Comparative studies of gregariousness and social structure among seven feral *Macaca fuscata* groups. In A. B. Chiarelli & R. S. Corruccine (Eds.), *Primate behavior and sociobiology* (pp. 52–63). Berlin: Springer.

Laland, K. N., Odling-Smee, J., & Myles, S. (2010). How culture shaped the human genome: Bringing genetics and the human sciences together. *Nature Reviews Genetics, 11*, 137–148.

Nakagawa, N. (1989). Feeding strategies of Japanese monkeys against the deterioration of habitat quality. *Primates, 30*, 1–16.

Nakagawa, N. (1990). Decisions on time allocation to different food patches by Japanese monkeys (*Macaca fuscata*). *Primates, 31*, 459–468.

Nakagawa, N. (1999). *Ecology of feeding rate: Feeding strategies in non-human primates.* Kyoto: Kyoto University Press [in Japanese].

Nakagawa, N. (2010). Intraspecific differences in social structures of the Japanese macaques: A revival of lost legacy by updated knowledge and perspective. In N. Nakagawa, M. Nakamichi, & H. Sugiura (Eds.), *The Japanese macaques* (pp. 271–290). Tokyo: Springer.

Nakagawa, N., Nakamichi, M., & Yamada, K. (2011). Report on the questionnaire for infrequently-observed behaviors in Japanese macaques. *Primate Research, 27*, 111–125 [in Japanese with English summary].

Nakamura, M. (2002). Grooming-hand-clasp in Mahale M group chimpanzees: Implications for culture in social behaviours. In C. Boesch, G. Hohmann, & G. Marchant (Eds.), *Behavioral Diversity in Chimpanzees and Bobobos* (pp. 71–83). New York: Cambridge University Press.

Nakamura, M., McGrew, W. C., Marchant, L. F., & Nishida, T. (2000). Social scratch: Another custom in wild chimpanzees? *Primates, 41*, 237–248.

Nakamura, M., & Nishida, T. (2006). Subtle behavioral variation in wild chimpanzees, with special reference to Imanishi's of *kaluchua. Primates, 47*, 35–42.

Nakamura, M., & Uehara, S. (2004). Proximate factors of different types of grooming hand-clasp in Mahale chimpanzees: Implications for chimpanzee social customs. *Current Anthropology, 45*, 108–114.

Nishida, T. (1973). The ant-gathering behavior by the use of tools among wild chimpanzees of the Mahale Mountains. *Journal of Human Evolution, 2*, 357–370.

Nishida, T., Mitani, J. C., & Wattts, D. P. (2004). Variable grooming behaviors in wild chimpanzees. *Folia Primatologica, 75*, 31–36.

Nishie, H. (2011). Natural history of *Camponotus* ant-fishing by the M group chimpanzees at the Mahale Mountains National Park, Tanzania. *Primates, 52*, 329–342.

Perry, S., Baker, M., Fedigan, L., Gros-Louis, J., Jack, K., MacKinnon, K. C., Manson, J. H., Panger, M., Pyle, K., & Rose, L. (2003). Social conventions in wild white-faced capuchin monkeys. *Current Anthropology, 44*, 241–268.

Sapolsky, R. M. (2006). Social cultures among nonhuman primates. *Current Anthropology, 47*, 641–656.

Sapolsky, R. M., & Share, L. J. (2004). A pacific culture among wild baboons: Its emergence and transmission. *PLoS Biology, 2*, 534–541.

Sugiyama, Y., & Koman, J. (1979). Tool-using and making behavior in wild chimpanzees at Bossou, Guinea. *Primates, 20*, 513–524.

Thierry, B. (2000). Covariation of conflict management patterns across macaque species. In A. Aureli & F. B. M. de Wall (Eds.), *Natural conflict resolution* (pp. 106–128). Berkeley: University of Califolnia Press.

Tishkoff, S. A., Reed, F. A., Ranciaro, A., Voight, B. F., Babbitt, C. C., Silverman, J. S., Powell, K., Mortensen, H. M., Hirbo, J. B., Osman, M., Ibrahim, M., Omar, S. A., Lema, F., Nyambo, T. B., Ghori, J., Bumpstead, S., Pritchard, J. K., Wray, G. A., & Deloukas, P. (2007). Convergent adaptation of human lactase persistence in Africa and Europe. *Nature Genetics, 39*, 31–40.

Todt, E., & Mikami, M. (2012). Dialogs: Going around disaster area in Tohoku. *Kan: History, Environment, Civilization, 48*, 142–151 [in Japanese].

Triandis, H. C. (1995). *Individualism and collectivism (new directions in social psychology).* Boulder: Westview Press.

Whiten, A., Goodall, J., McGrew, W. C., Nishida, T., Reynolds, V., Sugiyama, Y., Tutin, C. E. G., Wrangham, R. W., & Boesch, C. (1999). Cultures in chimpanzees. *Nature, 399*, 682–685.

Yamamoto, S., Yamakoshi, G., Humle, T., & Matsuzawa, T. (2008). Invention and modification of a new tool use behavior: Ant-fishing in trees by a wild chimpanzee (*Pan troglodytes verus*) at Bossou, Guinea. *American Journal of Primatology, 70*, 699–702.

Zhang, P., & Watanabe, K. (2007). Extra-large cluster formation by Japanese macaques (*Macaca fuscata*) on Shodoshima Island, Central Japan, and related factors. *American Journal of Primatology, 69*, 1119–1130.

Chapter 2
Primate Self-medication and the Treatment of Parasite Infection

Abstract Parasites cause a variety of diseases that affect the behavior and reproductive fitness of an individual. The study of animal self-medication as a science is relatively new. To date, research has classified health maintenance and self-medicative behaviors into four levels: (1) optimal avoidance or reduction of disease transmission; (2) the dietary selection of items with a preventative or health maintenance affect; (3) ingestion of a substance for the curative treatment of a disease or the symptoms thereof; and (4) external application of a substance to the body for the treatment or control of disease bearing insects. Of any species studied thus far, chimpanzees have provided the most details for level 2 and 3 behaviors, exemplified by such behaviors as bitter pith chewing and whole leaf swallowing behaviors, used in response to parasite infection. Both of these behaviors are behavioral adaptations that temporarily reduce parasite levels in the host.

Keywords *Aspilia* • Chimpanzees • Disease control • Ethnomedicine • Health maintenance • Host–parasite relationships • Pharmacology • *Vernonia amygdalina*

2.1 The Field of Primate Self-medication

It can be said that the study of primate self-medication began with the publication of a paper in 1983 by Richard Wrangham and Toshisada Nishida, two chimpanzee researchers then based at Harvard University and Tokyo University, respectively (Wrangham and Nishida 1983). They described a curious habit of chimpanzees at two East African research sites, Gombe and Mahale, in western Tanzania. They found folded, undigested, leaves from species of the genus *Aspillia*, in the dung and observed that chimpanzees would sometimes swallow these leaves whole without chewing them early in the morning after leaving their sleeping nests. Due to the unlikely nutritional value of such behavior, they proposed the possibility of some medicinal function, but provided no definitive evidence for a likely function or disease responsible. From the late 1980s onward, I intensified efforts to understand the factors causing illness and the function of various self-medicative behaviors in chimpanzees at these two sites. Today the topic of self-medication is rapidly growing, with a wide array of evidence not only from these two sites, but also from a growing number of different chimpanzee populations, for other great apes and monkeys, and indeed for many other animal species (e.g. Engel 2002; Huffman 1997, 2001, 2011).

The study of animal self-medication is devoted to understanding ways that animals respond to various threats to their health and provides a biological basis for the exploration of novel plant medicines and new ways they can potentially be used for the betterment of human health. Regardless of the species being studied, there are four basic requirements for demonstrating self-medication: (1) identify the disease or symptom(s) being treated; (2) distinguish the use of a therapeutic agent from that of everyday food items and or "medicinal foods"; (3) demonstrate a positive change in health condition following self-medicative behavior, and; (4) provide evidence for plant activity and or direct pharmacological analysis of compounds extracted from these plants. This chapter reviews the current state of our understanding about self-medication in primates and discusses this in the context of other animal species also being studied. Ultimately, the goal of this research is to shine light onto the non-human primate origins of our own medicinal behavior, an activity previously considered uniquely human, like culture was also once thought to be. By knowing where the ability to heal ourselves came from, we have the opportunity to take this information and apply it to finding ways of improving our health in the future to come.

2.2 Primate Self-medication and the Parasite Predicament

A hypothesis now being tested from the study of great ape self medication is that certain behaviors aid in the control of intestinal parasites and/or provide relief from related gastrointestinal upset.

At the individual level, self-medication may simply be driven by one's desire to be healthy and overcome discomfort. So far, the majority of evidence for self-medication in animals is about ways in which they deal with parasite infections. Parasites play a central role in the lives of wild animals since their existence is closely intertwined with that of parasites and pathogens. The ability of the individual, and by extension the species to defend itself against life-threatening disease provides a significant adaptive advantage and is thus predicted to occur throughout the animal kingdom. Much work exists describing the diversity of host–parasite relationships in the animal kingdom. Some parasitic infections seem to go unnoticed by the hosts. In other cases, when the health and survival of host is threatened by parasite infection, it is in the interest of the host to actively respond to alleviate discomfort.

2.3 Behavioral Strategies of Health Maintenance and Parasite Control

Given our current level of understanding, response to illness (parasite induced or otherwise), either as health maintenance or direct self-medicative behavior, can be classified into five levels: (1) "sick behaviors" (lethargy, depression, anorexia, reduction in grooming, behavioral fever, basking behavior (Kluger et al. 1975; Hart 1988); (2) *behavioral avoidance or reduction of the possibility for disease transmission* (avoidance of feces, contaminated food, water, substrates); (3) *dietary selection of items with a preventative or health maintenance effect* (items eaten routinely in small amounts or on a limited basis); (4) *ingestion of a substance for the curative treatment of a disease or the symptoms thereof* (use of toxic or biologically active items at low frequency or in small amounts, having little or no nutritional value), and; (5) *application of a substance to the body or a living space for the treatment or control of vectors or external health condition.* Behaviors to be focused on here fall within levels 3 and 4. In general terms, level 3 includes passive dietary prophylaxis or the consumption of "medicinal foods" as health maintenance behavior (Huffman 1997; Huffman et al. 1998). Emphasis is put on the passive nature of this category, as prophylactic treatment would imply intentionality and an understanding of both the cause and prevention of illness. To date, this has not been demonstrated in animals. Level 4 includes therapeutic behaviors such as the ingestion of pharmacologically active, non-dietary substances. This necessarily requires some level of awareness of wellness and discomfort and the ability to respond with behaviors that bring about positive change in one's condition.

Evidence for self-medication at these two levels has been found across the animal kingdom, with homologous behaviors appearing in phylogenetically distant taxa (e.g. swallowing of whole leaves for the expulsion of tapeworms and nodule worms in chimpanzees, gorillas, bonobos, brown bear, snow geese, civets, Japanese macaques, gibbons, wooly-bear caterpillars etc. see below). The diversity of potentially self-medicating species suggested thus far reflects the common need to prevent, suppress or cure parasite related disease. Current evidence suggest that for insects, self-medication may be operated by innate mechanisms (Bernays and Singer 2005), while for some higher vertebrates like chimpanzees, important aspects of self-medication, such as what plant species and when and how to ingest it, appear to be acquired and transmitted from generation to generation via socially biased learning and maintained in the group as culture (Huffman and Hirata 2004; Huffman et al. 2010).

2.4 Dietary Selection or Disease Prevention?

2.4.1 Medicinal Foods

Selecting a proper diet is important for energy, growth, general maintenance, and reproduction. Traditionally, feeding strategies are based upon finding a balance between the acquisition of essential nutritional elements such as carbohydrates, fats,

proteins and vitamins and the avoidance of the negative impact of secondary metabolites produced, which protect plants from over predation by reducing palatability
and/or digestibility for the many herbivore species (insect, vertebrates) that prey
upon them. However, this has not prevented some animals to actually benefit from
these plant-defense secondary compounds. The wooly-bear caterpillars of the tiger
moth (*Platyprepia virginalis*) protect themselves from the fatal effects of tachinid
parasitoid wasp larvae (*Thelaira americana*) infection by changing their diet from
innocuous lupine (*Lupinus arboreus*) to the toxic alkaloid abundant hemlock
(*Conium maculatum*) if they become infected. The alkaloids do not kill the developing parasite within the caterpillar's body, but allow the caterpillar to survive the
infection. Studies of such tri-trophic level interactions (insect host, plant, parasite)
are needed to deepen our understanding of the foundations for the evolution of general health maintenance and self-medication in the higher vertebrates (Bernays and
Singer 2005).

Earlier, I introduced the concept of "medicinal foods" to the primatological literature, and this has added an extra element of disease prevention based on the
consumption of plants with bioactive properties (Huffman 1997). This term is bor-
rowed from the human ethnopharmacological literature. Etkin (1996) and Etkin and
Ross (1983) reported that among the Hausa peoples of Nigeria, 30% of the wild
plant species identified and used by them as food, were also used as medicine.
Interestingly, 89% of the species were used by these people to treat symptoms of
malaria, caused by *Plasmodium* spp. Parasites. These plants were all used in a
dietary context. A variety of food items found in the diet of non-human primates
suggest that they too may actually be benefiting from small periodic doses of these
plant secondary metabolites (e.g. Krief et al. 2006; MacIntosh and Huffman 2010;
Pebsworth et al. 2006).

2.4.2 Nutrient Poor Items

Some potential medicinal foods found in the diets of many primate species include
bark and wood items. Bark and wood are by nature highly fibrous, heavily lignified,
sometimes toxic, relatively indigestible and nutrient poor (Huffman 1997). While
the list of plant species whose bark is ingested by primates is long, little is actually
known about the contribution of bark to the diet and general health. Chimpanzees
and gorillas infrequently ingest the bark and wood of several plant species. The bark
of *Pycnanthus angolensis* (Welw.) Warb. (Myristicaceae) ingested by chimpanzees
at Mahale in western Tanzania is used by West Africans as a purgative, laxative,
digestive tonic, emetic and reliever of toothaches. Chimpanzees at Gombe in western Tanzania occasionally eat the bark of *Entada abyssinica* (Mimosaceae). In
Ghana, the bark is used for diarrhea and as an emetic. The bark of *Gongronema lati-
folium* (Asclepiadaceae) occasionally eaten by chimpanzees at Bossou in West
Africa is extremely bitter, and some West Africans use the stems as a purge for
colic, stomach pains and symptoms connected with intestinal parasite infection.

The wood of several species ingested by gorillas and chimpanzees have also been shown to contain significant amounts of sodium, which may be lacking in the normal diet (Huffman 1997).

2.4.3 Hallucinogens and Stimulants

Stories of insects, birds and mammals under the influence of fermented fruit or other plant material are widespread in the wildlife lore and literature around the world, but few examples have been documented scientifically. One example from Africa stands out and is perhaps the best scientifically documented hallucinogen associated with use by animals. The plant is *Tabernanthe iboga* (Apocynaceae), a shrub first described in 1889 (Dubois 1955; Harrison 1968).

Today, *T. iboga* is used in religious rituals in Cameroon. Indigenous forest dwelling peoples reportedly discovered the hallucinogenic properties of this plant by watching gorillas, wild boars and porcupines digging up and eating the roots, afterwards going into a wild frenzy, running around as if fleeing from some frightening creature. The hallucinogenic affect of this plant is supported by pharmacological tests carried out on domestic animals in the laboratory in the early 1900s and later replicated in the 1950s. Today, the plant is under investigation in the US as an alternative for methadone, a standard treatment for drug addicts.

A review of all reports on gorilla diet found many well known medicinal plants with well-documented stimulatory, cardiotonic and hallucinogenic properties (Cousins and Huffman 2002). Gorillas utilize many species of Kola (*Cola*) (Sterculiaceae) particularly for their seeds. The seeds of *C. pachycarpa* for example contain such natural products as caffeine and theobromine. Kola seeds are low in protein, suggesting that gorillas obtain primarily caffeine from the seeds.

Gorillas ranging up in the high altitude regions (1,100–3,200 m above sea level) on the slopes of the Virunga volcano range feed on species of giant *Lobelia* (Campanlaceae) and *Senecio* (Compositae), both of which are important plants in ethnomedicine. All members of the genus *Lobelia* contain alkaloids (iobeline, iobelanidine and norlobelanidine) which can have a long lasting (up to 15 min) stimulatory effect upon the body if eaten. Higher doses can have narcotic effects. Gorillas reportedly feed on *L. giberroa* and *L. wallastonii* only occasionally (Cousins and Huffman 2002).

2.4.4 Antibiotic Properties

Many primates feed on figs, and many species have been shown to have anti-parasitic properties. Clinical trials on both humans and non-human animals have shown that some fig species are effective against the nematodes *Ascaris* and *Trichuris*. The active ingredient, ficin, a proteolytic enzyme, is present in all fig trees. Concentrations as low as 0.05% of the latex taken from *F. glabrata* have been shown to destroy the outer cutical of Ascarid worms and cause other lethal changes to the parasite's body.

Pith and fruit of species of the wild ginger family are also frequently eaten by chimpanzees, bonobos and gorillas (Huffman 1997). A detailed survey of the literature on *Afromomum* ginger species eaten by gorillas found significant bactericidal activities against a number of serious bacterial pathogens including *Escheria coli, Pseudomonas aeruginosa, Bacillus subtilis, and Proteus vulgaris*. Fungicidal activities were also found to inhibit *Candida albicans, Trichophyton mentagrophytes, Aspergillus niger*, and species of *Cladasporium* cladosporiodes (Cousins and Huffman 2002).

2.5   Therapeutic Self-medicative Behavior in Great Apes

2.5.1   Great Ape Parasites

Two therapeutic behaviors that have received the greatest amount of attention are leaf swallowing and bitter pith chewing (Fig. 2.1a, b). To date, two parasites have been associated with both bitter pith chewing and leaf swallowing; the nodular worm (*Oesophagostomum stephanastomum* and a primate tapeworm (*Bertiella studeri*) (Huffman 1997; Huffman and Caton 2001; Wrangham 1995). The nodular worm, 2–3 cm in length as adults, reinfect chimpanzees at the beginning of the rainy season, and it is at this time therapeutic self-medicative behaviors are practiced most frequently (Fig. 2.1). Nodular worms are one of the most hazardous parasite species in the great apes. Repeated infection, common in the wild, can cause secondary bacterial infection, diarrhea, severe abdominal pain, weight loss, and weakness, which can result in death. The ill effects of the tapeworm are not well documented, but abdominal pain is presumed to be associated with heavy infections (Wrangham 1995).

2.5.2   Bitter Pith Chewing, a Chemical Mode of Parasite Control

Bitter pith chewing is proposed to aid in the control of nodule worm infection, via pharmacological action, and relief from gastrointestinal upset. I first proposed the medicinal value of bitter pith chewing for chimpanzees from detailed behavioral observations, and parasitological and phytochemical analyses of patently ill chimpanzees' eating *Vernonia amygdalina* (Compositae), and recovering from their symptoms within 20 h (Huffman and Seifu 1989; Huffman et al. 1993).

When ingesting the pith from young shoots of *V. amygdalina*, chimpanzees meticulously remove the outer bark and leaves to chew on the exposed pith, from which they extract the extremely bitter juice (Fig. 2.1a). Only a relatively small amount is eaten at any one time, making any nutritional benefits insignificant. In total, $(5–120) \times 1$ cm are ingested. The entire process takes anywhere from less than 1 to 8 min, and it appears to be a one-dose treatment. Individuals have not been seen to eat more of the bitter pith again within the same day or even within the same week, likely due to the plant's toxicity. Infants of ill mothers, have been observed to taste small amounts of the pith discarded by their mothers, but adults close by rarely

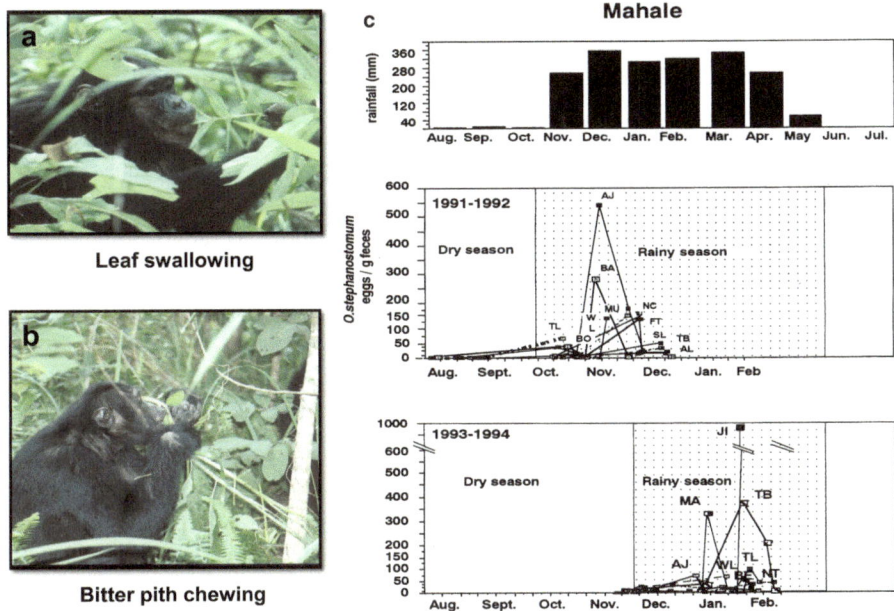

Fig. 2.1 Leaf swallowing (**a**) and bitter pith chewing (**b**), are two self-medicative behaviors in chimpanzees. At Mahale these two behaviors are most commonly observed during the rainy season months when the intensity of *Oesophagostomum stephanostomum* infections, measured by the number of eggs per gram feces, are at their highest (**c**)

eat the pith too, despite its availability. Even though it can be found year round, use of the bitter pith by chimpanzees is highly seasonal and rare, occurring most frequently during the rainy season peak in nodule worm infections—as expected for a therapeutic treatment.

Typically, a noticeable drop in appetite, malaise, diarrhea or constipation and high levels of nodule worm infection characterizes the state of health of individuals at the time of bitter pith chewing. Visible recovery from these symptoms can occur within 20–24 h of pith chewing (Huffman 1997).

For numerous African ethnic groups, a concoction made from *V. amygdalina* leaves or bark is prescribed treatment for malarial fever, schistosomiasis, amoebic dysentery, several other intestinal parasites and stomachaches. The noted recovery time of 20–24 h after bitter-pith chewing in chimpanzees is comparable to that of local human inhabitants of Mahale, Tanzania, the WaTongwe, who use cold concoctions of the plant's leaves as a treatment for parasites, diarrhea, malarial fever and stomach upset (Huffman et al. 1996). Phytochemical analysis has revealed the presence of two major classes of bioactive compounds; four sesquiterpene lactones and seven new stigmastane-type steroid glucosides (and two freely occurring aglycones of these glucosides) (Koshimizu et al. 1994).

The sesquiterpene lactones are recognized for their antiparasitic (nematode, ameba), antitumor, and antibiotic properties. From crude methanol extracts of the leaves, inhibition of tumor promotion and immunosuppressive activities have been demonstrated. In vitro tests on the antischistosomal activity of the pith's most abundant steroid glucoside (vernonioside B1) and sesquiterpene lactone (vernodaline) showed significant inhibition of movement of all adult stage parasites and of adult females' egg-laying capacity (Ohigashi et al. 1994).

Overall, the evidence from behavioral, parasitological, pharmacological and ethnomedicinal observations strongly support the hypothesis that bitter pith chewing is a therapeutic form of self-medication stimulated by, and controlling nodule worm infection.

2.5.3 Leaf-Swallowing, a Physical Mode of Parasite Control

Leaf-swallowing behavior in the African great apes was first reported for chimpanzees at Gombe and Mahale in 1983 (Wrangham and Nishida 1983). As of September 2010, leaf-swallowing behavior has been observed in 16 populations of chimpanzees (*Pan troglodytes schweinfurthii, P.t. troglodytes, P.t. verus*), bonobo (*P. paniscus*) and eastern lowland gorilla (*Gorilla gorilla graueri*) across Africa (Fig. 2.2). The leaf surface of the 40 plant species used, share the common property of being rough-surfaced, covered in stiff hairs, called trichomes. These trichomes are made of silicates that are difficult to digest and can also be very abrasive.

Leaf-swallowing has been shown to reduce nodule worm infection, and possibly relieve pain caused by tapeworm infection, by expelling these parasites. The mode of parasite control is a physical mechanism by means of a self-induced increase in gut motility, which acts as a purge. The behavior itself is quite distinct. The distal halves of these leaves are selected one at a time, folded by tongue and palate as they are slowly pulled into the mouth and then individually swallowed whole (Fig. 2.1). The roughness of the leaf surface makes it difficult to swallow, so folding them with the tongue and palate before swallowing is probably necessary, and is responsible for their exiting totally undigested (Huffman and Caton 2001).

An individual may swallow anywhere from 1 to 100 leaves in one sitting. Unlike bitter pith chewing, an individual may swallow leaves more than once in a day and over several days in a row. Chimpanzees swallow leaves within the first few hours after leaving their sleeping nests before the first meal and or on a relatively empty stomach.

2.6 Future Directions of Self-medication Research for the Health of Humans

This research is an example of how the disciplines of animal behavior, pharmacognosy and ethnomedicine can deepen our understanding about the behavior of animals. A practical application of the study of animal self-medication for human

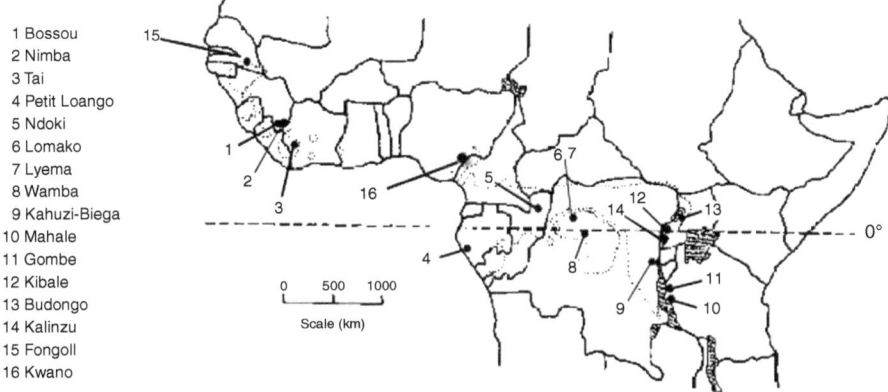

1 Bossou
2 Nimba
3 Tai
4 Petit Loango
5 Ndoki
6 Lomako
7 Lyema
8 Wamba
9 Kahuzi-Biega
10 Mahale
11 Gombe
12 Kibale
13 Budongo
14 Kalinzu
15 Fongoll
16 Kwano

Fig. 2.2 Distribution of leaf swallowing behavior across the entire distribution of chimpanzees, bonobos and gorillas. As of 2010, the behavior has been observed at 16 different study sites where great ape research is being conducted

health is to provide important leads to future sources of natural medicine for the treatment of parasite disease. In addition to this, a closer look into the ways that animals use these plants may provide new insights into new strategies for treating parasite disease in humans.

The next important step to realize this goal is to conduct in vivo tests to determine direct anthelmintic efficacy in a wide range of parasites, using a number of different host species. Differences in metabolism, drug pharmacokinetics etc. between monogastric and ruminant livestock necessitates further screening activities in the latter class of animals. Protocols for testing herbal preparations against nematode parasites of ruminants, using sheep as the model system and in vitro/in vivo model testing of plant extract and raw plant material as foliage against such significant diseases as schistosomiasis and leishmaniasis using captive primate species as models are also desired.

This multidisciplinary approach to research, where biological activity of novel, plant derived compounds acquired from the study of animal behavior and ethnomedicine is assessed against a whole range of parasite species found across a wide range of hosts, maximizes the chance of success. At the same time the author recognizes the importance of preserving the intellectual property rights of the regions/countries to any new discoveries derived from indigenous plant material. A prime objective of this extension to self-medication research should be to integrate our results into local health care and livestock management systems so that locally available plants can be properly used to the benefit of all in parts of the world where western medicine is often too expensive to buy and impractical to use due to the acute exposure to parasites.

References

Bernays, E. A., & Singer, M. S. (2005). Taste alteration and endoparasites. *Nature, 436*, 476.

Cousins, D., & Huffman, M. A. (2002). Medicinal properties in the diet of gorillas—an ethnopharmacological evaluation. *African Studies Monographs, 23*, 65–89.

Dubois, L. (1955). *Tabernanthe iboga* Baillon. *Bulletin Agricole du. Congo Belgique, 46*, 805–829.

Engel, C. (2002). *Wild health*. Boston: Houghton Mifflin Company.

Etkin, N. L. (1996). Medicinal cuisines: diet and ethnopharmacology. *International Journal of Pharmacognosy, 34*(5), 313–326.

Etkin, N. L., & Ross, P. J. (1983). Malaria, medicine, and meals: Plant use among the Hausa and its impact on disease. In L. Romanucci-Ross, D. E. Moerman, & L. R. Tancredi (Eds.), *The anthropology of medicine: From culture to method* (pp. 231–259). New York: Praeger.

Harrison, G. P. (1968). *Tabernanthe iboga*: an African narcotic plant of social importance. *Economic Botany, 23*, 174–184.

Hart, B. L. (1988). Biological basis of the behavior of sick animals. *Neuroscience and Biobehavioral Reviews, 12*, 123–137.

Huffman, M. A. (1997). Current evidence for self-medication in primates: a multidisciplinary perspective. *Yearbook of Physical Anthropology, 40*, 171–200.

Huffman, M. A. (2001). Self-medicative behavior in the African great apes—an evolutionary perspective into the origins of human traditional medicine. *Bioscience, 51*, 651–661.

Huffman, M. A. (2011). Primate self-medication. In C. Campbell, A. Fuentes, K. MacKinnon, M. Panger, & S. Bearder (Eds.), *Primates in perspective* (2nd ed., pp. 563–573). Oxford: University of Oxford Press.

Huffman, M. A., & Caton, J. M. (2001). Self-induced increase of gut motility and the control of parasitic infections in wild chimpanzees. *International Journal of Primatology, 22*, 329–346.

Huffman, M. A., Gotoh, S., Izutsu, D., Koshimizu, K., & Kalunde, M. S. (1993). Further observations on the use of the medicinal plant, *Vernonia amygdalina* (Del) by a wild chimpanzee, its possible effect on parasite load, and its phytochemistry. *African Study Monographs, 14*, 227–240.

Huffman, M. A., & Hirata, S. (2004). An experimental study of leaf swallowing in captive chimpanzees—insights into the origin of a self-medicative behavior and the role of social learning. *Primates, 45*, 113–118.

Huffman, M. A., Ohigashi, H., Kawanaka, M., Page, J. E., Kirby, G. C., Gasquet, M., et al. (1998). African great ape self-medication: A new paradigm for treating parasite disease with natural medicines? In Y. Ebizuka (Ed.), *Towards natural medicine research in the 21st century* (pp. 113–123). Amsterdam: Elsevier Science.

Huffman, M. A., Page, J. E., Sukhdeo, M. V. K., Gotoh, S., Kalunde, M. S., Chandrasiri, T., et al. (1996). Leaf-swallowing by chimpanzees, a behavioral adaptation for the control of strongyle nematode infections. *International Journal of Primatololgy, 17*, 475–503.

Huffman, M. A., & Seifu, M. (1989). Observations on the illness and consumption of a possibly medicinal plant *Vernonia amygdalina* by a wild chimpanzee in the Mahale Mountains, Tanzania. *Primates, 30*, 51–63.

Huffman, M. A., Spiezio, C., Sgaravatti, A., & Leca, J.-B. (2010). Option biased learning involved in the acquisition and transmission of leaf swallowing behavior in chimpanzees (*Pan troglodytes*)? *Animal Cognition, 13*, 871–880.

Kluger, M. J., Ringler, D. H., & Anver, M. R. (1975). Fever and survival. *Science, 188*, 166–168.

Koshimizu, K., Ohigashi, H., & Huffman, M. A. (1994). Use of *Vernonia amygdalina* by wild chimpanzees; possible roles of its bitter and related constituents. *Physiology and Behavior, 5*, 1209–1216.

Krief, S., Huffman, M. A., Sévenet, T., Guillot, J., Hladik, C.-M., Grellier, P., et al. (2006). Bioactive properties of plant species ingested by chimpanzees (*Pan troglodytes schweinfurthii*) in the Kibale National Park, Uganda. *American Journal of Primatology, 68*, 51–71.

MacIntosh, A. J. J., & Huffman, M. A. (2010). Towards understanding the role of diet in host-parasite interactions in the case of Japanese macaques. In F. Nakagawa, M. Nakamichi, & H. Sugiura (Eds.), *The Japanese macaques* (pp. 323–344). Tokyo: Springer.

Ohigashi, H., Huffman, M. A., Izutsu, D., Koshimizu, K., Kawanaka, M., Sugiyama, H., et al. (1994). Toward the chemical ecology of medicinal plant use in chimpanzees: the case of *Vernonia amygdalina*, a plant used by wild chimpanzees possibly for parasite-related diseases. *Journal of Chemical Ecology, 20*, 541–553.

Pebsworth, P., Krief, S., & Huffman, M. A. (2006). The role of diet in self-medication among chimpanzees in the Sonso and Kanyawara communitites, Uganda. In N. E. Newton-Fisher, H. Notman, V. Reynolds, & J. Paterson (Eds.), *Primates of Western Uganda* (pp. 105–133). New York: Springer.

Wrangham, R. W. (1995). Relationship of chimpanzee leaf-swallowing to a tapeworm infection. *American Journal of Primatology, 37*, 297–303.

Wrangham, R. W., & Nishida, T. (1983). *Aspilia* spp. leaves: a puzzle in the feeding behavior of wild chimpanzees. *Primates, 24*, 276–282.

Chapter 3
From Genes to the Mind: Comparative Genomics and Cognitive Science Elucidating Aspects of the Apes That Make Us Human

Abstract Primate Research Institute of Kyoto University has a long history in comparative cognitive science, represented by the "Ai Project." The researchers in this field attempt to answer how and why the human mind evolved, which components of the mind are shared between humans and chimpanzees, and which are uniquely human. Because the human mind and brain functions (a source of mind) are products of evolution, their evolutionary trajectories should be affected by evolutionary constraints, and such signatures should remain in the genome. In the field of genome science, rapid advances in high-throughput technologies have enabled the availability of whole-genome sequences at the individual level. A rapidly advancing discipline related to cognitive science and genomics, termed cognitive genomics, aims to uncover how the genome contributes to the structure and function of the human brain and mind. In particular, cognitive genomics investigates how genomic products, such as the transcriptome, proteome, methylome, and metabolome, relate to brain and cognitive functions in a temporal–spatial manner. We would like to expand this perspective from an evolutionary point of view. Here, we review recent progress in the field of comparative cognitive science, various types of primate genome databases, and chimpanzee whole-genome analysis conducted by our group.

Keywords Chimpanzee • Cognitive science • Comparative genomics • Gene • Genome database • Mind

M.A. Huffman et al., *Monkeys, Apes, and Humans: Primatology in Japan*, SpringerBriefs in Biology, DOI 10.1007/978-4-431-54153-0_3, © The Author(s) 2013

3.1   Exploring the Chimpanzee Mind: Thirty Five Years of Comparative-Cognitive Studies of Chimpanzees at the Primate Research Institute (PRI)

3.1.1   Comparative Cognitive Science

Psychology usually attempts to answer how the human mind works. However, the scope of comparative cognitive science is much broader. In this field, researchers investigate how and why the human mind evolved. Researchers in this field have been trying to address these attractive but difficult questions. There are a number of methods available to answer these questions using different subjects. Among the different subjects, we focus on the chimpanzee, the closest evolutionary neighbor of humans. By exploring various aspects of the chimpanzee mind using cognitive science methodologies, psychology, and ethology, we attempted to understand which components of the mind are shared by humans and chimpanzees and which are uniquely human. Based on these components of the mind, we will be able to reconstruct the most probable evolutionary history of the mind and further infer the factors affecting the evolution of the human mind. For this purpose, 35 years ago studies of the chimpanzee mind were initiated at the PRI, Kyoto University by Kiyoko Murofushi, Toshio Asano, Tetsuya Kojima, Tetsuro Matsuzawa, and their colleagues.

3.1.2   Teaching Visual Symbols to Chimpanzees: The Initiation of the "Ai Project"

In 1969, PRI acquired one female infant chimpanzee, who spent her first 10 years without another chimpanzee mate. PRI started its comparative cognitive studies in 1978 when a second chimpanzee, Ai, was acquired and began visual artificial language training (Matsuzawa 1991, 2003, 2006; Asano et al.1982).

During the 1960s and 1970s, there were increasing numbers of research projects designed to teach "language" to great apes. Researchers had trained chimpanzees, gorillas, orangutans, and bonobos to use sign language or visual symbols (Gardner and Gardner 1969; Miles (1999); Patterson and Linden 1981; Premack 1976; Rumbaugh 1977; Savage-Rumbaugh and Lewin 1994; Terrace 1979). These projects reported how many words apes could learn and how many sentences they produced and comprehended. However, the conclusion to date is that chimpanzees cannot learn the "syntax" of human-like language. Murofushi and the colleagues thus focused on how chimpanzees learn "words." They prepared systematic visual symbols called "lexigrams" composed of geometric components. By combining two or three components, one complex stimulus had one specific "meaning." Using a matching to sample paradigm, chimpanzees learned the meanings of these "words" (Fig. 3.1). The experiments initially included three chimpanzees. Although two of

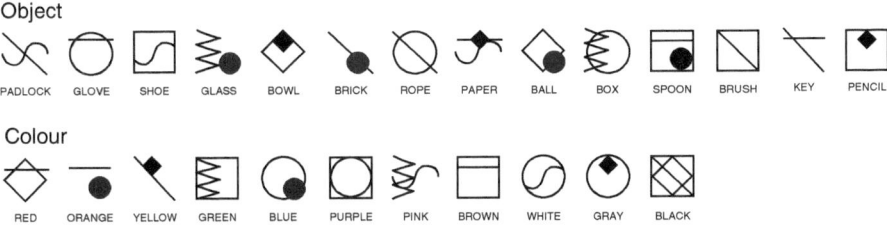

Fig. 3.1 Lexigrams learned by the chimpanzee Ai (from Matsuzawa 1985)

the three chimpanzees learned the names of colors (Asano et al. 1982); Matsuzawa 1991), additional experimental projects included Ai alone. Later, this project became known as the "Ai Project" (Matsuzawa 1991).

Ai learned the names of 11 colors, 14 objects, and 6 foods (Fujita and Matsuzawa 1990; Matsuzawa 1991). She also learned to match Arabic numerals from 1 to 9 to corresponding numerosities. Furthermore, she spontaneously fixed the "word-order" to describe multiple numbers of colored objects (such as five red pencils). Ai dominantly described the color or object name first and the number last (such as "RED," "PENCIL," "5"; Matsuzawa 1985). She was also trained to order the words in a manner predetermined by the experimenter (Murofushi et al. 1988). When she was presented a video clip showing person A approaching person B, she could describe this event as "A" "APPROACH" "B."

What had Ai learned during these word-learning experiments? Are the words she learned really the same as those in human language? To address these questions, one important property of words called stimulus equivalence was tested (Sidman and Tailby 1982). When humans learn a word in the direction from the word to the object, we can readily choose the object in response to the spoken word without any explicit training. Such bidirectionality between the word and the meaning is referred to as "symmetry," which is one of the important features for stimulus equivalence. If a stimulus–stimulus relation has the following features, then these stimuli can be equivalent, or we can say that the relation is an equivalence relation: reflexivity (A=A), symmetry (if A=B, then B=A), and transitivity (if A=B and B=C, then A=C). In humans, irrespective of being typically or atypically developed, stimulus equivalence (symmetry, in particular) is readily established. In chimpanzees, irrespective of a history of language training, the emergence of symmetry without any explicit training is very difficult. Yamamoto and Asano (1995) tested whether Ai exhibited symmetry and found no evidence of immediate and spontaneous emergence of symmetry, but they did observe a gradual development of symmetrical stimulus relations between symbol and meaning; repeated bidirectional training had a facilitative effect on the emergence of symmetry (see also Tomonaga 2008).

Our conclusion to date concerning the linguistic ability of chimpanzees is that chimpanzees can learn the association between arbitrary visual symbols and objects and simple word order, but these abilities are not identical to human linguistic competence.

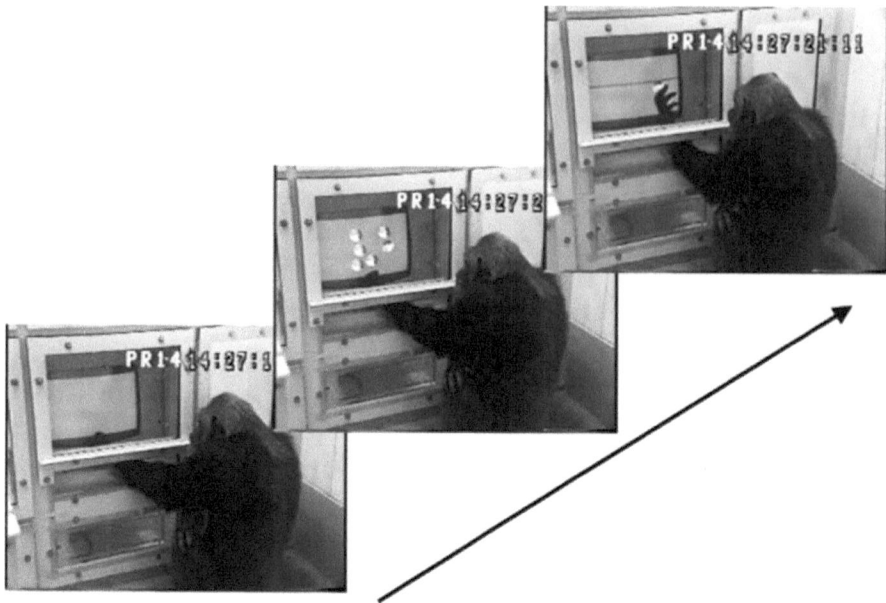

Fig. 3.2 The chimpanzee Chloe performing the visual search task for face orientation (from Tomonaga 2007a, b)

After 30 years, biolinguistic studies are now being revived (Hauser et al. 2002). Unfortunately, the (positive) contributions to these movements from the "ape-language" studies were seemingly low (e.g., Terrace 2011). We need to investigate linguistic competence in great apes from different perspectives.

3.1.3 How Do Chimpanzees See the World?

After the worldwide decline in the prevalence of ape-language studies, comparative cognitive science turned to more basic aspects of chimpanzee cognition in the late 1980s. One of these trends is comparative studies of visual cognition. For example, Tomonaga (2001a) used the visual search and related tasks that are frequently applied to the study of human visual cognition to examine the basic properties of visual cognition in chimpanzees. In the visual search task, the participants are required to search the display for one target amongst several distractor stimuli and respond to the target.

Tomonaga and colleagues found that many aspects of visual cognition in chimpanzees are similar to those in humans. For example, the shift of attention in a chimpanzee is triggered by the abrupt visual onset of peripheral stimuli (Tomonaga 1997). Chimpanzees exhibit a clear inversion effect for face perception (Tomonaga 1999a, b, 2007a; Fig. 3.2), and they can perceive biological motion patterns composed of movements of point lights (Tomonaga 2001b). Chimpanzees also exhibited improved discrimi-

Fig. 3.3 Examples of hierarchically organized stimuli (*left*) and Pendesa performing the visual search task using these stimuli (Fagot and Tomonaga 1999)

nation of geometric stimuli when specific contexts that enhanced the configurations were added (configural superiority effect, Goto et al. 2012).

We also found relevant differences between humans and chimpanzees. One of the most striking differences is found in perceptual organization. For example, Fagot and Tomonaga (1999) tested the chimpanzee perception of hierarchically organized patterns shown in Fig. 3.3. When humans see these patterns, we readily focus on the global pattern composed of local features (circle composed of squares), whereas chimpanzees do not display such a bias. Matsuno and Tomonaga (2006) found that chimpanzees struggled to integrate the coherently moving dots as a "surface."

We also examined higher-order visual cognition in chimpanzees, in particular, the recognition of social stimuli such as faces. As described previously, chimpanzees process faces in the same manner as humans: they perceive the faces not based on their local features (e.g., eyes, nose, and mouth) but on the spatial configuration of these features. There are, however, some critical differences in social cognition between humans and chimpanzees. One of these differences was found in the perception of another individual's gaze direction. In humans, the gaze direction of another individual reflexively triggers the shift of visuospatial attention toward the direction of the gaze (Friesen and Kingstone 1998). This shift of attention occurs very quickly (within 100 ms) and is very robust against top-down manipulation of the reliability of the gaze direction. In contrast to humans, chimpanzees exhibited rather different patterns. Similar to humans, the attention of chimpanzees is readily captured by the face (Tomonaga 2007b; Tomonaga and Imura 2009a). However, the disengagement and shift of attention are different between these species. For example, Kano and Tomonaga (Kano et al. 2011) found significant differences in the performance of gap-overlap experiments between these species. In the gap-overlap

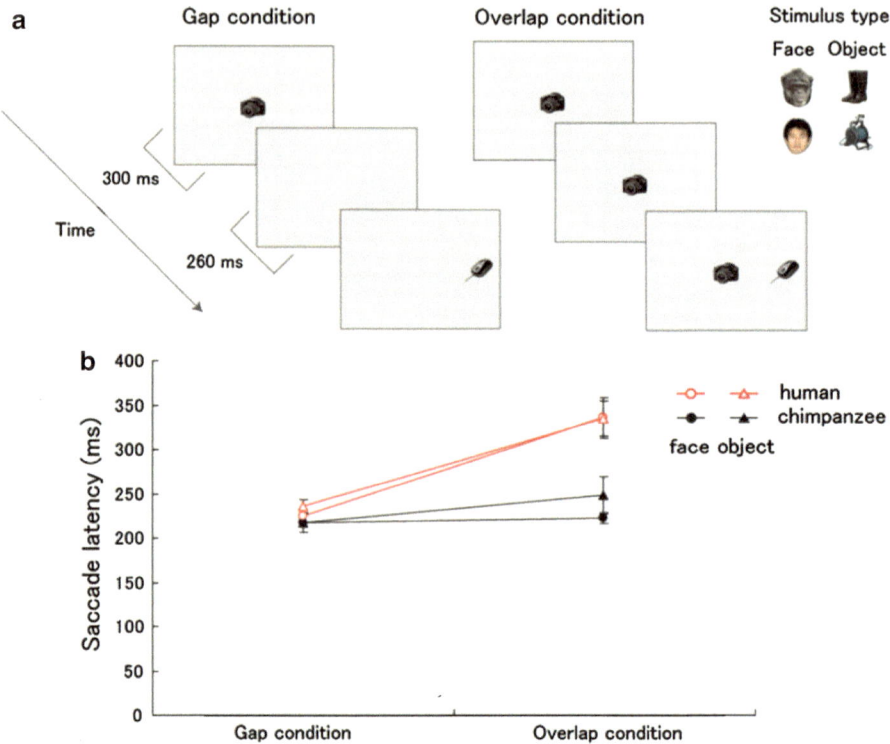

Fig. 3.4 Schematic diagrams of the gap-overlap paradigm (**a**) and the mean saccadic latency for each species and condition (from Kano and Tomonaga 2011)

paradigm, the participants are initially presented a stimulus in the center of the monitor. After focusing on this stimulus, another stimulus appears in periphery.

In humans, the response to the peripheral stimulus was faster when this stimulus appeared after termination of the central stimulus (gap condition) than when the central stimulus remained on the screen during the presentation of the peripheral stimulus (overlap condition). In contrast to humans, chimpanzees displayed no differences in performance between the gap and overlap conditions (Fig. 3.4). In summary, chimpanzee attention is captured by the abrupt peripheral onset of social or nonsocial stimuli (Tomonaga and Imura 2009b), but this attention is easily disengaged from those stimuli irrespective of the stimulus content (Kano and Tomonaga 2011). Furthermore, Tomonaga and colleagues (Tomonaga 2007a; Tomonaga and Imura 2009a) found that a shift of attention toward the gaze direction was also different between chimpanzees and humans. In these experiments, chimpanzees were required to discriminate the letters or detect the letter presented on the left and/or right. Before presenting the stimulus (or stimulus pair), social cues (eye gaze, head orientation, and pointing gestures) signaled the forthcoming target position. As a result, chimpanzees exhibited completely different performance compared with humans. Chimpanzees exhibited a significant cueing effect (facilitated performance

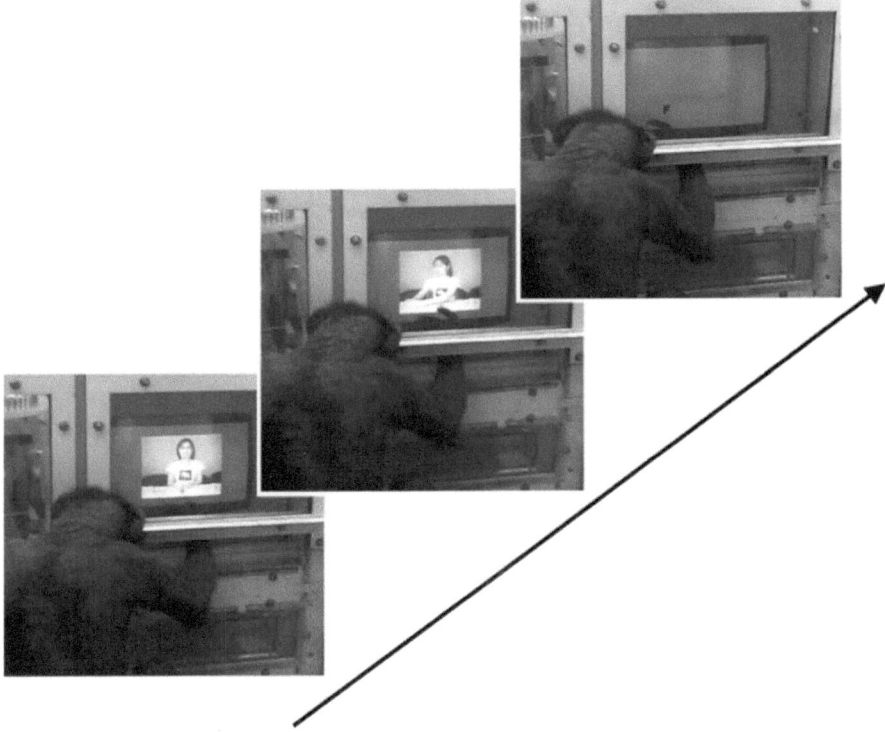

Fig. 3.5 The chimpanzee Pendesa performing the social cueing task (from Tomonaga and Imura 2009b)

when the cue and the target location were congruent and vice versa) only when the gaze cue was highly reliable. In addition, the cueing effect required more than 100 ms to activate (Fig. 3.5). Related to this finding, Hattori et al. (2010) also reported that chimpanzee attention was only shifted by conspecific social cues, whereas human attention was shifted by both human and chimpanzee cues (Fig. 3.6). These results can be discussed with respect to the qualitative differences in social interactions (Tomonaga 2010; Tomonaga et al. 2004) and anatomical differences in eye morphology (Kobayashi and Kohshima 1997) between these species.

3.1.4   Comparative–Cognitive–Developmental Perspective

The human mind is undoubtedly a product of evolution. In addition, the development of the human mind also should be a result of evolution. It is not sufficient to compare the minds of adult humans and adult chimpanzees alone, it is necessary to compare the entire developmental trajectories of the minds of these two species.

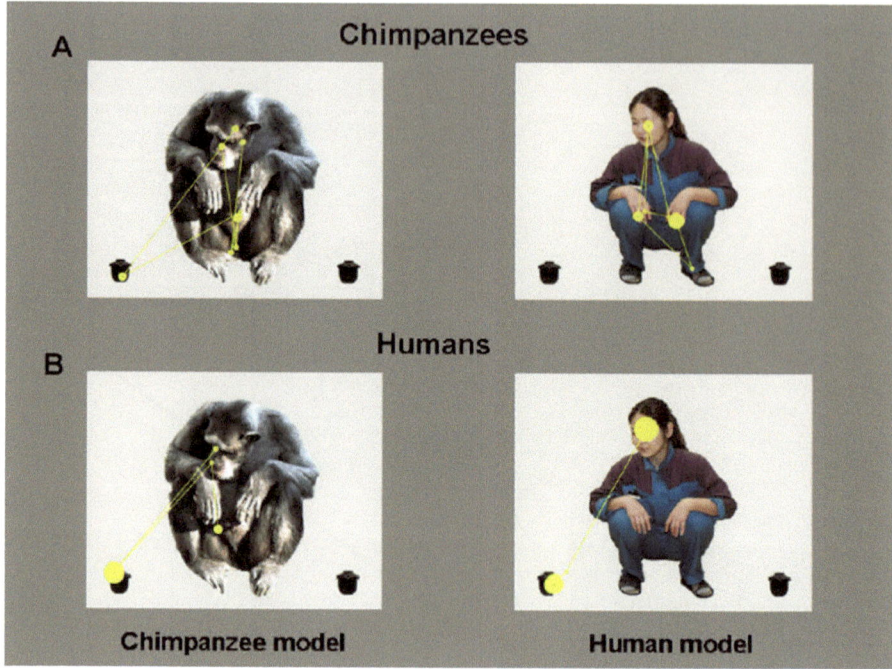

Fig. 3.6 Gazing patterns of chimpanzees and humans when viewing social signals (from Hattori et al. 2010)

Beginning in 2000, PRI started a new stage of the Ai Project that is a longitudinal study of the cognitive development of chimpanzees (Matsuzawa et al. 2006; Tomonaga et al. 2003). There have been several studies on the development of the chimpanzee mind (e.g., Hayes 1951; Kellogg and Kellogg 1933). In all these studies, however, chimpanzee infants were not raised by their own mothers. The researchers studied the development of the minds of chimpanzees raised in human cognitive–social environments by human caregivers. Ape-language projects were one of the extreme examples of this approach. In contrast, we studied the longitudinal changes in behavior and cognition of chimpanzee infants raised by their own mothers living with other community members in enriched environments (Matsuzawa 2006). We investigated three mother–infant chimpanzee pairs and examined various aspects of the cognitive ability of infants before their birth (e.g., Kawai et al. 2004), during the neonatal phase, and through their infancy (Fig. 3.7).

Research topics were relatively varied. Among these topics, in particular, we focused on the development of social cognition. For example, we tested their ability to recognize their mother's face (Myowa-Yamakoshi et al. 2005) and found that at 1 month of age, the infant chimpanzees could discriminate between the faces of their mothers and those of other chimpanzees. Interestingly, this discrimination apparently disappeared at 2 months of age (Fig. 3.8). During this period, they became more interested in social stimuli such as faces and attempted to communicate with others in the face-to-face setting by using a mutual gaze (Bard et al. 2005) and

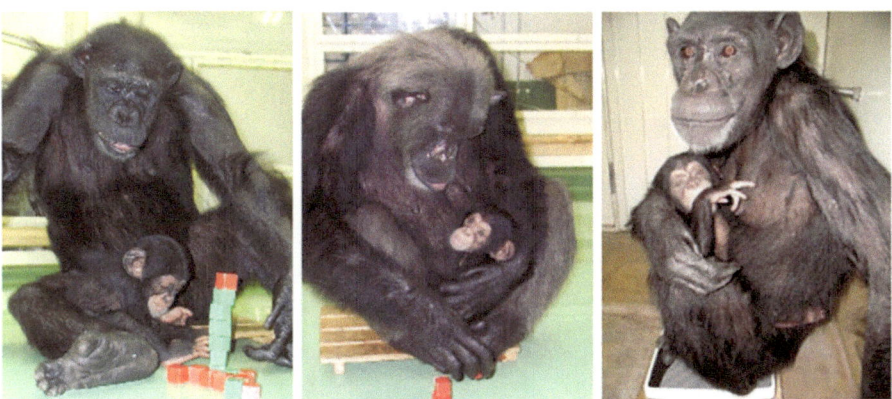

Fig. 3.7 Three chimpanzee mother–infant pairs. Infants were all born in 2000. From *left* to *right*; Ai (mother) and Ayumu (infant), Chloe and Cleo, and Pan and Pal. Photo courtesy: Primate Research Institute, Kyoto University

Fig. 3.8 The infant chimpanzee Cleo looking at her mother's face mounted on the CCD camera (photograph: Mainichi Shinbun)

"social smile" (Mizuno et al. 2006). At 4–5 months of age, the infants began to move alone and started to explore the environment. At this phase, the mother and infant frequently exchanged gazes (Okamoto-Barth et al. 2007), suggesting the emergence of attachment. In summary, there appeared to be no critical differences in social–cognitive development during early infancy between chimpanzees and humans. At the later phase of infancy (from 9 months to 2 years), however, chimpanzee and human infants exhibited different developmental trajectories regarding social cognition. One of the most striking differences was the absence of "joint attention" or triadic interactions among the infant, mother, and objects (Tomonaga et al. 2004). In humans, a qualitative change concerning social communication

Fig. 3.9 The chimpanzee Ayumu was examined for his gazing patterns of pictures using a non-restrained infrared eye tracking system. Photo courtesy: Primate Research Institute, Kyoto University

occurs at approximately 9 months of age (Tomasello 1999); at this age, human infants start to engage in triadic exchanges with others. Their interactions involve both objects and people, resulting in the formation of a referential triangle of infant, adult, and the object upon which they share attention. This type of interaction is the basis for the later development of "theory of mind" in humans. In contrast, chimpanzee infants exhibit no signs of such developmental changes. In fact, they can follow another individual's gaze direction (Okamoto et al. 2002), but this gaze-following ability did not facilitate the emergence of (human-like) triadic interactions. Nevertheless, the lack of this ability in chimpanzees does not mean they are "less intelligent" compared with humans. We should consider why humans and not chimpanzees acquired this social–cognitive ability during evolution. We should further explore what (social–cognitive) factors in the environment were responsible for the evolution of social interactions uniquely observed in humans.

3.1.5 Step Toward the Next Decade of the Twenty-First Century

In this section, we briefly describe the history of comparative cognitive studies of the chimpanzee mind at PRI. We have accumulated various types of evidence. Because of technological advances, we are now examining new aspects of the chimpanzee mind, such as eye-movement patterns (Kano and Tomonaga 2009; Fig. 3.9) and the sense of self-agency (Kaneko and Tomonaga 2011). Furthermore, we are exploring the possibilities of collaborative research projects with researchers in other fields such as genomics. The relationship between individual differences in cognition and behavior and those in individual genomes may provide insight into the rise of "comparative cognitive genomics" (e.g., Konopka and Geschwind 2010).

3.2   Primate Genome Database

3.2.1   Introduction

As mentioned in Sect. 3.2, Kyoto University has a long history in the field of primate biology, including fieldwork and the comparative cognitive study named the "Ai Project." Based on these primate studies, as a part of the global COE project to form a strategic base for "evolution and biodiversity research," we constructed the "Primate Genome Database" in 2008 (http://gcoe.biol.sci.kyoto-u.ac.jp/pgdb/). Using this database, we plan to reconsider these fieldworks and cognitive achievements from genomic points of view. We are analyzing the genomic sequences of 14 chimpanzees living in PRI, Kyoto University. On this website, the data for some genes are presented with the physiology, behavior, and personality of each individual. This website may provide a unique opportunity for users to consider what primates, including human beings, are.

The website is constructed on the basis of three concepts. The first concept is, "Humans are also a primate species." Some people consider humans to be exceptional; however, *Homo sapiens* are also primates. On accessing this website, you can notice that humans are classified as primates. The second concept is "individuality." The primate research of Kyoto University is always conducted at the individual level. We believe that recognizing all primates, including humans, as equals is imperative in understanding their personalities. Therefore, we constructed our database at the individual level. We hope that general readers will consider each chimpanzee as an individual and enjoy viewing the database at the individual level. The third concept is "from genome to field." To understand the evolution of primates, including humans, fieldwork with laboratory experimentation is required. Thus, we constructed this website to let users access both fieldwork and laboratory experiments. We hope that this database will help users form a new image of primates. In this section, we introduce the details of the database according to the webpage sequences.

3.2.2   Geographical Information

In this webpage, the geographical birthplace of each chimpanzee is shown on a Google map. On clicking the location tab, users are linked to Africa and presented videos showing the life of chimpanzees in the wild. If you click "Present place," users can see the locations of PRI residents in Kyoto University and zoom in to see their living spaces.

3.2.3   Family Tree

Almost all the chimpanzees housed at PRI are related to each other. On clicking the links for individual chimpanzees, users can see the name, sex, species, subspecies, birthday, present location, birth location, father, mother, registry number, features, characters, individual history, photos, videos, biometry, chromosome, and microarray data of each chimpanzee.

3.2.4   Gallery

Many videos of chimpanzees submitted to YouTube are linked in this webpage. These videos were recorded from 2001 to 2009. Users can see the unique behaviors of chimpanzees as described in Sect. 3.2.

3.2.5   Genotype Comparison

The majority of the genome data are focused on bitter taste receptor (TAS2R or T2R), olfactory receptors (ORs), and neural-related genes, the genotypes of which are reported to be related to personality in humans.

3.2.5.1   TAS2R

Mammals can perceive and distinguish five basic taste qualities, namely sweet, bitter, sour, salty, and umami, the taste of glutamate (for review, see Chandrashekar et al. 2006; Sugawara and Imai 2012). Of these modalities, bitter sensitivity has a particularly important role. Many naturally poisonous substances taste bitter to humans, and virtually all animals exhibit an aversive response to such tastants (Hilliard et al. 2004; Ueno et al. 2004; Chandrashekar et al. 2006), suggesting that bitter transduction evolved as a key defense mechanism against the ingestion of harmful substances. Several studies have revealed that bitter taste in mammals is mediated by TAS2Rs, which belong to the large family of seven-transmembrane G protein-coupled receptors, expressed in specialized taste bud cells of the tongue and palate epithelium (Adler et al. 2000; Chandrashekar et al. 2000; Matsunami et al. 2000; Bufe et al. 2002; Conte et al. 2003; Shi et al. 2003; Mueller et al. 2005; Behrens and Meyerhof, 2006, 2009; Behrens et al. 2007). The coding regions of *TAS2R*s are ~900 base pairs (bp) long and are not interrupted by introns. Genomic analyses using comparisons with reference sequences revealed the repertoires of *TAS2R* in various vertebrates (Go 2006; Shi and Zhang 2006).

Inter-specific comparisons of sequences in each *TAS2R* from one or two individuals of different primates revealed that some pseudogenes in one species appear to be functional genes in the other species or vice versa (Parry et al. 2004; Fischer et al. 2005; Go et al. 2005). Most differences, however, are at the single base pair level (Parry et al. 2004), and the repertoires of *TAS2R*s in humans and non-human primates are not remarkably different (Parry et al. 2004; Fischer et al. 2005; Go et al. 2005). These facts prompted us to examine intra-species variations of *TAS2R*s in chimpanzees to better understand the biological significance of bitter perception (Sugawara et al. 2011). To date, polymorphisms of *TAS2R*s in non-human primates have rarely been examined excluding *TAS2R38* in chimpanzees (Wooding et al. 2006). Based on these motivations, in this database, we present the sequences of TAS2Rs of each individual chimpanzee at PRI.

As a result of sequence analyses of TAS2Rs, we found several single nucleotide polymorphisms (SNPs) even in the individuals living at PRI. For example, TAS2R1, the receptor for bitter substances such as chloramphenicol, humulone, diphenylthiourea, picrotoxinin, and yohimbine, exhibited a 1-bp deletion at position 739 in some individuals. Chimpanzees Ayumu, Akira, Ai, Mari, and Chloe have a heterozygous genotype for this position. These facts indicate the existence of a segregating pseudogene for this gene in chimpanzees because the deletion causes a frame shift in the middle of this gene. This suggests that some individuals cannot recognize bitter compounds specifically recognized by TAS2R1. This is similar to the case of TAS2R38, in which start codon of ATG was mutated to AGG and this gene could not be translated to a normal seven-transmembrane-domain protein (Wooding et al. 2006; Sugawara et al. 2011; Imai in press). This segregating pseudogene is one of the molecular basis of differences in taste sensitivity between individuals.

Typical SNPs are found in almost all TAS2Rs. TAS2R4, which recognized bitter substances such as denatonium benzoate and PROP, contains three SNPs. One of them is a non-synonymous mutation producing the codon CGG or TGG, which code for Arg and Trp, respectively. Because this mutation occurs at amino acid position 123, the surface of the intracellular compartment, the mutation cause a functional difference in TAS2R4 and the bitter taste perception of foods recognized by this TAS2R. In TAS2R16, which recognizes glucopyranosides such as salicin, polymorphisms exist at amino acid position 126. Chimpanzees Reiko, Reo, and Creo have a heterozygous genotype of TGG and TGT at position 384, whereas the others have the homozygous genotype of TGG. Because TGG and TGT are codons for Trp and Cys, respectively, this mutation causes some functional differences between the wild-type and mutant proteins. In fact, our analysis of expressed proteins confirmed the functional significance of this position (Imai et al. 2012).

Interestingly, some specific mutations were observed in Pendesa, a Western, Central, and Eastern hybrid chimpanzee. For example, the codon at position 522 of TASR1 in Pendesa is TCA instead of TCC. Although this SNP is a synonymous polymorphism, several specific mutations exist in the TAS2Rs and ORs of Pendesa. These facts suggest differences in the genotypes of chimpanzee subspecies, some of which could be related to differences in the phenotypes of the subspecies. It should be noted that the medical use of bitter plants is observed in specific fields. The sequence difference in bitter taste receptors could be one of the molecular basis of regional differences in the use of medical plants (Huffman, see also chapter 2).

3.2.5.2   OR

Sequence analyses revealed individual and regional differences in ORs as well as TAS2Rs. ORs comprise the largest family of G protein-coupled receptors, and each species has a specific OR gene set in its genome (Go and Niimura 2008; Matsui et al. 2010; Niimura 2012). In addition, individual differences exist within the species (Adipietro et al. 2012). These results suggest that chimpanzees living in different areas would have different repertoires of ORs and different senses of smell. This trend is also observed in TAS2Rs, especially among the bitter taste receptors. These data demonstrated that environmental differences in living regions are related to the differences in the repertoires of olfactory and taste receptors.

It is necessary to investigate the function of these receptors to link genotypes to environmental-specific phenotypes. Although most of the ligands for most ORs are unknown, some have been rapidly deorphanized by Dr. Matsunami's group at Duke University (Adipietro et al. 2012). For example, Or7D4 senses androstenone and androstegienon, which are a sex hormone and a candidate pheromone, respectively. Interestingly, the sequences of some individuals are different from those of others, and it is possible that the sense of smell is different among different individuals. Annotation of other ligand and receptor pairs may explain specific behavior-related senses of smell. For this purpose, we demonstrated that the sequences of some ORs that are specific for chimpanzees and human do not have orthologs (Go and Niimura 2008). Any researchers with knowledge of ligands for these receptors are welcome to collaborate in the investigation of their phenotypes.

3.2.5.3   Neural-Related Gene

This section presents polymorphisms primarily in the repeat length of neural-related genes. You can link these genotypes to the phenotypic personalities shown on the website.

First, the length of the polymorphic region of the neurotransmitter dopamine receptor D4 gene (*DRD4*) is shown (Shimada et al. 2004). Polymorphisms have been observed in the second intron of *DRD4* in humans and apes, and some are related to the personalities of humans.

Second, the numbers of repeats of the polymorphic region of the dopamine transporter gene (*DAT1*, *SLC6A3*) are shown. DAT is an integral membrane protein that removes dopamine from the synaptic cleft and deposits it into surrounding cells, thus terminating the signal of the neurotransmitter. Dopamine underlies several aspects of cognition, including reward, and DAT facilitates regulation of that signal.

The repeat number of the polymorphic region of the serotonin transporter gene (*5HTTP*, solute carrier family 6 (neurotransmitter transporter, serotonin), member 4, *SLC6A4*) is shown. 5HTTP moves serotonin from the synaptic cleft back into the synaptic boutons. Thus, it terminates the effects of serotonin and simultaneously enables its reuse by the presynaptic neuron. A polymorphism was observed in the intronic region that was reported to affect the personality of humans. This polymorphism is found in three human alleles: 9-, 10-, and 12-bp repeats. Regarding this gene in chimpanzees, 18- and 19-bp repeats are mainly observed. Interestingly, only Pendesa has a 23-bp repeat, whereas Chloe has a 25-bp repeat.

The repeat number of the polymorphic region in the gene promoter region of monoamine oxidase (*MAOA*), which metabolizes neurotransmitters (Inoue-Murayama et al. 2006), is also listed. A variable number of tandem repeats (VNTR) polymorphism based on a 30-bp unit has been reported in the promoter region of human *MAOA*. Human VNTRs have been demonstrated to affect transcriptional activity, and some reports suggest that VNTR polymorphisms are associated with psychoneurological disorders. In a human neuroblastoma cell line, most of the ape sequences that had a short repeat length (12 or 18 bp) exhibited higher promoter activity than a human 3-repeat sequence with a 30-bp repeat length. However, an intra-specific difference dependent on the repeat number was not observed among the ape alleles examined. In case of Japanese macaques, the polymorphism exist but a consistent tendency was not observed with dominance rank (Inoue-Murayama et al. 2010). There would be a need to increase the number of samples and candidate genes, and personality study should be considered.

The repeat numbers of Gln and Gly polymorphic regions in the androgen receptor (AR) are also shown (Choong et al. 1998). AR receives androgen, which is essential for male reproductive development and virilization. Therefore, the polymorphism may affect reproductive behaviors or some related phenotypes.

The lengths of the polymorphic regions of estrogen receptors alpha (*Era*) and beta (*ERb*), in the gene promoter and intronic regions, respectively, are also shown.

Nonsynonymous polymorphisms in tryptophan hydroxylase 2 (*TPH2*), the gene involved in serotonin synthesis (Hong et al. 2011), are listed. It was reported that a polymorphism in *TPH2* is associated with chimpanzee neuroticism. In the brain, serotonin production is controlled by TPH2. Previous studies found that mutations on the TPH2 locus in humans were associated with depression, and studies of mice and rhesus macaques demonstrated that the TPH2 locus is related to aggressive behavior. Hong et al. (2011) reported a functional SNP in the form of an amino acid substitution, Q468R, in the chimpanzee TPH2 gene-coding region. They tested whether this SNP was associated with neuroticism in captive and wild-born chimpanzees living in Japan and Guinea, respectively. This study was the first to identify a genotype linked to a personality trait in chimpanzees.

We hope that the linkage between genes and behaviors are further elucidated in the near future. There would be an environmental factors needed to trigger genomic effects. For example, the mutation of Glu 86 in TAS2R16 in macaques caused the decrease in the sensitivity for glucopyranosides like salicin. This effect becomes significant in Japanese macaques which ingest bark of trees including bitter glucopyranosides in the winter season (Imai et al. 2012). Because wild primates live in the species specific niches, the environmental factors would take some parts in triggering the genomic effects in the behaviors.

3.2.6   Chromosome Image

The chromosome images can be observed for almost all of the individuals. The column "rDNA labeling" indicates the 18S rDNA, which exists as five pairs of signals in each image. Ag labeling indicates the previously active rDNAs, and DAPI staining indicates chromosome specific band-patterns with identified chromosome numbers.

3.2.7   Microarray

Microarray data are available for Ayumu, Akira, Pal, and Creo. The data were acquired for the total RNAs of lymphocytes based on the NimbleGen human microarray (Euk Expr 4x72K Catalog Arr (HG18 60 mer expr 4plex), based on HG18.36).

3.2.8   Personality Comparison

The scores concerning seven different characteristics of each individual are presented in this section. It is possible to compare the relationships of each characteristic among different individuals. We recommend moving from this page to the genomic information or videos page, where you can see their behaviors, we hope you enjoy exploring this website.

3.2.9   Genotype–Phenotype Relationship

Using this type of database, genotype–phenotype associations can be elucidated at the individual level. For TAS2R, the ligands have been more deorphanized than those for ORs by Dr. Meyerhof's group (Meyerhof et al. 2010). For example, TAS2R38 is the receptor for PTC, which is perceived as a bitter compound by some humans but is tasteless to others. The genotypes and phenotypes were linked recently for humans, chimpanzees, and Japanese macaques. Therefore, individual differences exist in the tastes of these ligands in many primate species including humans (Chiarelli 1963; Wooding et al. 2006; Suzuki et al. 2010). Primate researchers performed behavioral tests to confirm these phenotypes. First, Ueno et al. (2004) reported data on the individuals living at PRI. Second, Dr. Morimura and coworkers applied a more sophisticated application to the chimpanzees in Kumamoto Sanctuary, Kyoto University, based on the genotype shown on the website and in the literature (Morimura, Sugawara et al., unpublished results). The results are clear and suggest that the sensory receptors and neuronal enzymes are candidates for the model system of genotype–phenotype relationship.

3.3   Chimpanzee Genome Studies at an Individual Level at PRI

3.3.1   Advances in Genome Science in the Last Ten years

For more than 10 years, papers reporting the first draft of the human genome were simultaneously published in *Nature* by the Human Genome Project (International Human Genome Sequencing Consortium 2001) and *Science* by Celera Genomics (Venter et al. 2001). In addition to publishing the near-complete human genome sequence in 2004 (International Human Genome Sequencing Consortium 2004), many genome projects have been initiated and completed in the past decade. Genomes for organisms including the mouse (Mouse Genome Sequencing Consortium 2002), rat (Rat Genome Sequencing Project Consortium 2004), dog (Lindblad-Toh et al. 2005), chimpanzee (Chimpanzee Sequencing and Analysis Consortium 2005), rhesus macaque (Rhesus Macaque Genome Sequencing and Analysis Consortium 2007), marsupial (Mikkelsen et al. 2007), and most recently the Neanderthal (Green et al. 2010) have been reported. Regarding the genomes of primates, in addition to the chimpanzee and rhesus macaque, genomes for many other primates have emerged, such as the gorilla, orangutan, gibbon, baboon, marmoset, tarsier, galago, and lemur.

Recently, sequencing technology has been revolutionized, and the capacity for producing sequencing reads has been tremendously improved in recent years. This includes the production of a massive parallel sequencer, called the next-generation sequencer (NGS), which permits reading several tens of billion bases per run. The per-base speed of NGS sequencing has increased ~100,000-fold over the past decade, far outpacing Moore's law, an empirical rule regarding the rate of advancement of computer processing capabilities (e.g., processing speed, memory capacity). The current NGS machines can read ~300 billion bases in a week, compared to ~25,000 in 1990 and ~5 million in 2000, indicating that ~10,000-fold larger sequence data can be obtained by NGS than by the capillary-based Sanger sequencer. NGS technology, by enabling vast data generation, has provided a comprehensive picture of human genome variation. For example, the 1000 genome project, launched in 2008, is an international research consortium to establish the most detailed catalog of human genetic variation, and this project was designed to sequence the genomes of at least 1,000 anonymous participants from a number of different ethnic groups (http://www.1000genomes.org/). In 2010, the project finished its first pilot phase, and the details were published (The 1000 Genomes Project Consortium 2010). As of the end of 2011, the number of individuals with entirely sequenced genomes was predicted to exceed 10,000. The speed and cost of sequencing have also dramatically decreased. Whereas it took approximately 15 years and cost 300 million dollars to produce the first full human genome sequence, NGS technology now enables us to obtain an entire genome sequence within several days for a cost of <5,000 dollars.

The innovation of sequencing technologies has provided significant benefits to many aspects of medical genomics. In particular, many researchers now focus on deciphering the genomic and molecular mechanisms of a wide range of cancers by comparing genomic sequences between tumor cells and their counterparts (normal cells) and by detecting *de novo* somatic mutations that are specific to each tumor cell. These high-throughput characteristics can be applied to evolutionary and comparative genomic studies as well as human medical and population studies. Application of NGS technologies to the chimpanzees housed at PRI are briefly described in the following sections.

3.3.2   Application of NGS Technologies to Comparative Genomics Studies

Comparative genomics is the study of the relationships of genomic structure and function across different species and populations. In particular, we are conducting comparative genomics analysis by using non-human primates as an appropriate reference organism to humans. This is done because we need appropriate out-groups for a deeper and intrinsic understanding of "what makes us human." As mentioned above, extensive investigations have been performed to uncover human polymorphisms in the last decade. By contrast, data on sequence variations and polymorphisms in non-human primates are very limited. Therefore, the primate genome project was launched as a collaborative project among the Kyoto University global COE program, PRI of Kyoto University, and the National Institute of Genetics. Through the project, we focus on performing whole-genome sequencing in multiple chimpanzees by utilizing NGS technologies. It is well known that chimpanzees are the closest living relative to humans, and many distinguished psychological studies of chimpanzees, known as the "Ai Project," were conducted over a period of 35 years at PRI (see Sect. 3.2).

3.3.3   Chimpanzee Genomics at an Individual Level to Understand Their Variations

The term "personal genomics" originated from the medical and pharmacological fields of human genomics. Personal genomics focuses on sequencing and analyzing the genome of an individual. In personal genomic studies on a scale of hundreds or even thousands (such as the 1000 genome project), many types of variations, including single nucleotide variations (SNVs), variation of insertions and deletions (InDels), copy number variations (CNVs), inversions, and retroelement insertions, can be found among individuals and compared with the published literature to determine the likelihood of qualitative and quantitative phenotypic traits. These traits substantially include all the individual differences related to physical traits, disease risk, and even personality. Obtaining the entire genomic sequence of non-human primates as well as humans thus has the potential to link genomic variations to various types of qualitative and quantitative phenotypic traits. As a first step in an individual chimpanzee genomics, we selected a trio of chimpanzees: Akira (father), Ai (mother), and Ayumu (son) (Fig. 3.10).

Fig. 3.10 A chimpanzee family trio, Akira (father), Ai (mother), and Ayumu (son), used for the whole-genome sequencing

To perform whole-genome sequencing, high-quality DNA is needed, and DNA from each chimpanzee was collected from blood and extracted using a standard DNA extraction kit. After several quality control checks, a DNA library for NGS application was made for each individual. For sequencing, we used a massive parallel high-throughput sequencer, AB SOLiD4 (Life Technologies) (Fig. 3.11), through which several hundreds of million short reads (~50 bp) can be obtained per run, resulting a total read of ~100 gigabases (Gb) per run.

Due to the disadvantage of the short read length (~100 bp) of NGS technology, more sequence reads (ordinarily more than a 30-fold larger amount of sequence reads over the genome size, which is conventionally denoted as >30×) are needed to cover nearly the full length of the genomic sequence compared with the conventional capillary-based sequencer, in which most of the genome projects have sequenced at most sixfold larger than the genome size (~6×). In this study, we sequenced more than 45× (45 times) raw reads in the three individuals under the assumption of a chimpanzee genome size of 3.35 Gb and obtained 38–47× mapped or reference matched reads, which is equivalent to 130–160 Gb, after excluding unmapped reads. The statistics of whole-genome sequencing are summarized in Table 3.1.

All the 50-bp long sequence reads were mapped to the chimpanzee reference sequence (panTro2 in the UCSC genome database: http://genome.ucsc.edu/cgi-bin/hgGateway?db=panTro2). The reference whole-genome data were derived primarily from the donor Clint, a captive-born male Western chimpanzee (*Pan troglodytes verus*) from the Yerkes Primate Research Center in US, and the total sequence length of panTro2, excluding estimated gap sizes, is 2.91 Gb. Approximately three-fourths of the reads, more than 3 billion reads, can be mapped to the reference genome, covering ~98% (2.59–2.64 Gb) of the panTro2 genomic sequence (Table 3.1).

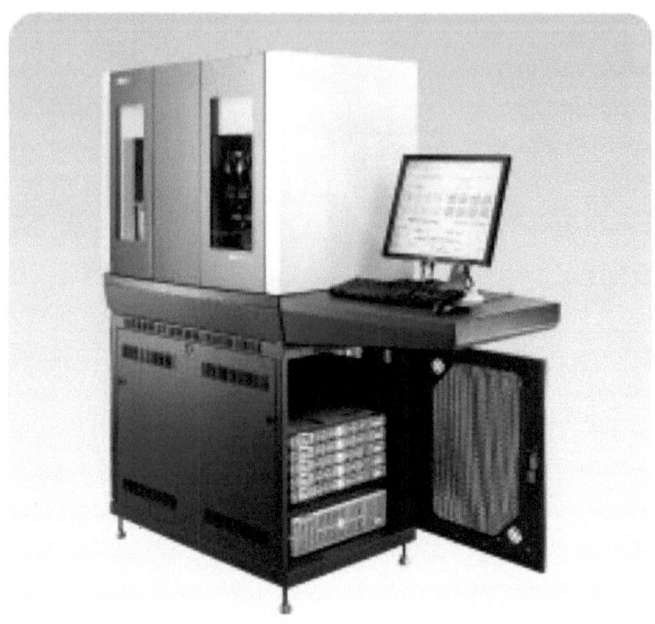

Fig. 3.11 The AB SOLiD4 next-generation sequencer

Table 3.1 Statistical summary of whole-genome sequencing in a chimpanzee family trio

Individual	Sequenced bases (raw)	Read depth (raw)[a]	No. of mapped read	Mapped read[a] (%)	Sequenced bases (mapped)	Read depth (mapped)[a]	Genome coverage[a]
Akira	159.8 Gb	47.7×	3,293,301,003	75.80	132.3 Gb	39.5×	97.83%
Ai	153.8 Gb	45.9×	3,135,655,725	73.30	129.0 Gb	38.5×	97.86%
Ayumu	179.6 Gb	53.6×	3,808,519,483	73.10	158.8 Gb	47.4×	98.61%

All reads are mapped to the chimpanzee reference sequence (panTro2)
[a]Assuming the chimpanzee reference genome size of 3.35 Gb

The ultimate goal of whole-genome sequencing is to uncover the causal links between genotypes (genome) and phenotypes (phenome) as mentioned previously. The first step toward achieving this goal is to catalog all types of individual variations in each genome. With this aim, we attempted to detect various types of variations using panTro2 as a reference sequence. As a result, 2.35–2.52 million SNVs and 192–251 thousand InDels were detected over the entire genome (Table 3.2). By using the number of SNVs, the nucleotide diversity for Akira, Ai, and Ayumu was 0.091%, 0.093%, and 0.096%, respectively. Previous studies of human personal genomics have revealed that the number of SNVs in the human genome exceeds three million, which is more than that found in the chimpanzee genome (Table 3.3), possibly because of a smaller effective population size for Western chimpanzees (~7,400) (Hey 2010) than for humans (~10,200) (Yu et al. 2004).

Table 3.2 Number and types of SNVs in a chimpanzee family trio

Individual	No. of SNVs[a]	p (%)[b]	No. of InDels[c]	No. of synonymous SNVs[d]	No. of nonsynonymous SNVs[e]	No. of stop-introducing SNVs[f]
Akira	2,347,892	0.0905	192,572	9,564	9,986	87
Ai	2,412,052	0.0933	225,471	9,469	9,798	79
Ayumu	2,522,357	0.0956	251,568	9,317	9,531	72

All reads were mapped to the chimpanzee reference sequence (panTro2) to identify variation
[a]Single nucleotide variations
[b]The nucleotide diversity was calculated from the number of SNVs in each individual divided by the total number of covered bases
[c]Insertions and deletions
[d]Single nucleotide variations within a protein coding gene that do not alter the amino acid sequence
[e]Single nucleotide variations within a protein coding gene that alter the amino acid sequence
[f]Single nucleotide variations within a protein coding gene that introduce a termination codon, resulting in impaired gene function

Table 3.3 Number of single nucleotide variations (SNVs) in the genomes of different individualls

Individual	Ethnic group	No. of SNVs
Craig Venter	Caucasian	3,213,401
James Watson	Caucasian	3,322,093
Yoruban (Nigerian)	Negroid	3,699,592
Han Chinese	Mongoloid	3,074,097
Seong-Jin Kim (Korean)	Mongoloid	3,439,107
NA18943 (Japanese)	Mongoloid	3,132,608

The SNVs found in each genome are from Craig Venter (Levy et al. 2007), James Watson (Wheeler et al. 2008), and individuals of Yoruban (Bentley et al. 2008), Han Chinese (Wang et al. 2008), Korean (Ahn et al. 2009), and Japanese ethnicity (Fujimoto et al. 2010)

Within protein-coding regions, SNVs are classified into three types of variations: synonymous SNVs, which are variations not altering the amino acid sequence and hence presumably having no effect on fitness (selectively neutral); nonsynonymous SNVs, which are variations altering the amino acid sequence and are subjected to the selection; and stop-introducing SNVs, which are variations introducing a termination codon and resulting in impaired gene function. Extensive and careful analysis revealed 9,317–9,564 synonymous SNVs, 9,531–9,986 nonsynonymous SNVs, and 72–87 stop codon-introducing SNVs in the chimpanzee trio (Table 3.2). These numbers are substantially similar to the numbers detected in the human trio whole-genome sequencing project, a pilot study of the 1000 genome project: 9,193–9,414 and 11,508–12,500 synonymous SNVs, 8,299–11,122 and 10,349–11,122 nonsynonymous SNVs, and 55–69 and 78–84 stop codon-introducing SNVs in Caucasian and Negroid trio families, respectively (The 1000 Genomes Project Consortium 2010).

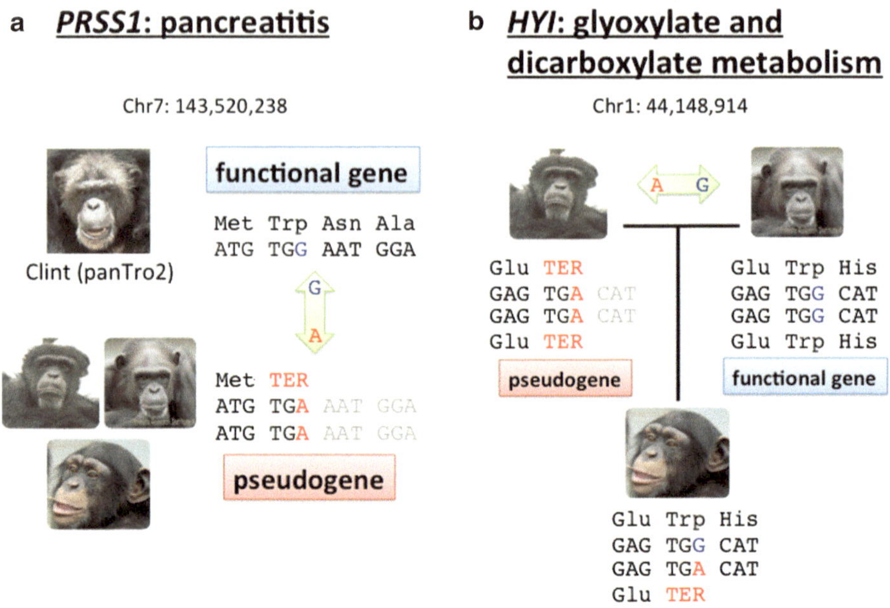

a _PRSS1_: pancreatitis

b _HYI_: glyoxylate and dicarboxylate metabolism

Chr7: 143,520,238

Chr1: 44,148,914

Clint (panTro2)

functional gene

Met Trp Asn Ala
ATG TGG AAT GGA

G
A

Met TER
ATG TGA AAT GGA
ATG TGA AAT GGA

pseudogene

A   G

Glu TER   Glu Trp His
GAG TGA CAT   GAG TGG CAT
GAG TGA CAT   GAG TGG CAT
Glu TER   Glu Trp His

pseudogene   functional gene

Glu Trp His
GAG TGG CAT
GAG TGA CAT
Glu TER

Fig. 3.12 Examples of stop codon-introducing single nucleotide variations (SNVs) in the chimpanzee family trio. (**a**) _PRSS1_ encodes a trypsinogen that is secreted by the pancreas and cleaved to its active form in the small intestine. It is known that mutations of this gene in humans are associated with hereditary pancreatitis. In the chimpanzee trio, all three individuals have a stop codon-introducing SNV (substitution from G to A) and are believed to carry the pseudogene. (**b**) _HYI_ is involved in glyoxylate and dicarboxylate metabolism. This gene is polymorphic in terms of function among the three individuals: Akira carries both pseudogenized alleles, Ai carries both functional alleles, and Ayumu carries one functional and one pseudogenized allele

Among SNVs, we focused on stop codon-introducing SNVs because this type of SNV possibly leads to loss of gene function due to the insertion of a stop codon into the amino acid sequence. As mentioned previously, we found 72–87 stop codon-introducing SNVs (Table 3.2) compared with the reference genomic sequence, implying that an approximate 100-gene variance in the total number of genes among individuals existed. Two examples of such SNVs are shown in Fig. 3.12. The first is _PRSS1_, which encodes a trypsinogen, a member of the trypsin family of serine proteases. This enzyme is secreted by the pancreas and cleaved to its active form in the small intestine. It targets peptide linkages involving the carboxyl group of lysine or arginine. In humans, mutations in this gene are associated with hereditary pancreatitis (Whitcomb et al. 1996; Gorry et al. 1997; Férec et al. 1999). In our chimpanzee trio, all three individuals have a stop codon-introducing SNV as a consequence of a G-to-A substitution, probably indicating the lack of function of this gene in Akira, Ai, and Ayumu (Fig. 3.12a). The second example is _HYI_, which is putatively involved in glyoxylate and dicarboxylate metabolism. Although the function of the gene is not well known, this gene was polymorphic in terms of function among the three chimpanzees (Fig. 3.12b). In particular, Ai has both functional alleles, but Akira has two pseudogenized alleles. As a corollary, Ayumu has one functional and one pseudogenized allele in his genome. These examples are good candidate genes for uncovering clues about individual phenotypic differences.

3.3.4 Further Perspective for Linking the Genome and Phenome

The field of personal genomics in non-human primates is in its infancy, along with the rapid spread of NGS technology. Within human genomics, personal genomics is developing very rapidly, primarily due to its potential applicability to medical and pharmacological fields. Several thousand or even tens of thousand human genomes are being entirely sequenced to identify sequence variants/mutations in causal genes that are related to human diseases, such as cancers and psychiatric disorders. In principle, the same methodology can be applied to the field of primatology to both find disease-causing genes and to better understand the personality or mind of primates. Fortunately, our institute has a long history of breeding a large number of monkeys (mainly Japanese and rhesus monkeys) as well as chimpanzees, and multimodal studies have been performed in many fields, including morphology, physiology, ethology, veterinary science, and psychology. In the near future, integrating the knowledge obtained from those fields into the field of genomics will facilitate a comprehensive understanding of genome-based human commonality and specificity from a primatology perspective.

References

Adipietro, K. A., Matsunami, H., & Zhuang, H. (2012). Functional evolution of primate odorant receptors. In H. Hiai, H. Imai, & Y. Go (Eds.), *Post-genome biology of primates* (pp. 63–78). Tokyo: Springer.

Adler, E., Hoon, M. A., Mueller, K. L., Chandrashekar, J., Ryba, N. J., & Zuker, C. S. (2000). A novel family of mammalian taste receptors. *Cell, 100*, 693–702.

Ahn, S. M., Kim, T.-H., Lee, S., Kim, D., Chang, H., Kim, S.-S., et al. (2009). The first Korean genome sequence and analysis: full genome sequencing for a socio-ethnic group. *Genome Research, 19*, 1622–1629.

Asano, T., Kojima, T., Matsuzawa, T., Kubota, K., & Murofushi, K. (1982). Object and color naming in chimpanzees *Pan troglodytes*. *Proceedings of the Japan Academy, 58B*, 118–122.

Bard, K. A., Myowa-Yamokoshi, M., Tomonaga, M., Tanaka, M., Costal, A., & Matsuzawa, T. (2005). Group differences in the mutual gaze of chimpanzees (*Pan troglodytes*). *Developmental Psychology, 41*, 616–624.

Behrens, M., Foerster, S., Staehler, F., Raguse, J. D., & Meyerhof, W. (2007). Gustatory expression pattern of the human TAS2R bitter receptor gene family reveals a heterogenous population of bitter responsive taste receptor cells. *The Journal of Neuroscience, 27*, 12630–12640.

Behrens, M., & Meyerhof, W. (2006). Bitter taste receptors and human bitter taste perception. *Cellular and Molecular Life Sciences, 63*, 1501–1509.

Behrens, M., & Meyerhof, W. (2009). Mammalian bitter taste perception. *Results and Problems in Cell Differentiation, 47*, 203–220.

Bentley, D. R., Balasubramanian, S., Swerdlow, H. P., Smith, G. P., Milton, J., Brown, C. G., et al. (2008). Accurate whole human genome sequencing using reversible terminator chemistry. *Nature, 456*, 53–59.

Bufe, B., Hofmann, T., Krautwurst, D., Raguse, J. D., & Meyerhof, W. (2002). The human TAS2R16 receptor mediates bitter taste in response to beta-glucopyranosides. *Nature Genetics, 32*, 397–401.

Chandrashekar, J., Hoon, M. A., Ryba, N. J., & Zuker, C. S. (2006). The receptors and cells for mammalian taste. *Nature, 444*, 288–294.

Chandrashekar, J., Mueller, K. L., Hoon, M. A., Adler, E., Feng, L., Guo, W., et al. (2000). T2Rs function as bitter taste receptors. *Cell, 100*, 703–711.

Chiarelli, B. (1963). Sensitivity to P.T.C (phenyl-thio-carbamide) in primates. *Folia Primatologica, 1*, 88–94.

Chimpanzee Sequencing and Analysis Consortium. (2005). Initial sequence of the chimpanzee genome and comparison with the human genome. *Nature, 437*, 69–87.

Choong, C. S., Kemppainen, J. A., & Wilson, E. M. (1998). Evolution of the primate androgen receptor: a structural basis for disease. *Journal of Molecular Evolution, 47*, 334–342.

Conte, C., Ebeling, M., Marcuz, A., Nef, P., & Amdres-Barquin, P. J. (2003). Evolutionary relationships of the Tas2r receptor gene families in mouse and human. *Physiological Genomics, 14*, 73–82.

Fagot, J., & Tomonaga, M. (1999). Global-local processing in humans (*Homo sapiens*) and chimpanzees (*Pan troglodytes*): Use of a visual search task with compound stimuli. *Journal of Comparative Psychology, 113*, 3–12.

Férec, C., Raguénès, O., Salomon, R., Roche, C., Bernard, J. P., Guillot, M., et al. (1999). Mutations in the cationic trypsinogen gene and evidence for genetic heterogeneity in hereditary pancreatitis. *Journal of Medical Genetics, 36*, 228–232.

Fischer, A., Gilad, Y., Man, O., & Pääbo, S. (2005). Evolution of bitter taste receptors in humans and apes. *Molecular Biology and Evolution, 22*, 432–436.

Friesen, C. K., & Kingstone, A. (1998). The eyes have it! reflexive orienting is triggered by nonpredictive gaze. *Psychonomic Bulletin and Review, 5*, 490–495.

Fujimoto, A., Nakagawa, H., Hosono, N., Nakano, K., Abe, T., Boroevich, K. A., et al. (2010). Whole-genome sequencing and comprehensive variant analysis of a Japanese individual using massively parallel sequencing. *Nature Genetics, 42*, 931–936.

Fujita, K., & Matsuzawa, T. (1990). Delayed figure reconstruction by a chimpanzee (*Pan troglodytes*) and humans (*Homo sapiens*). *Journal of Comparative Psychology, 104*, 345–351.

Gardner, R. A., & Gardner, B. T. (1969). Teaching sign language to a chimpanzee. *Science, 165*, 664–672.

Go, Y. (2006). Lineage-specific expansions and contractions of the bitter taste receptor gene repertoire in vertebrates. *Molecular and Biological Evolution, 23*, 964–972.

Go, Y., & Niimura, Y. (2008). Similar numbers but different repertoires of olfactory receptor genes in humans and chimpanzees. *Molecular Biology and Evolution, 25*, 1897–1907.

Go, Y., Satta, Y., Takenaka, O., & Takahata, N. (2005). Lineage-specific loss of function of bitter taste receptor genes in humans and nonhuman primates. *Genetics, 170*, 313–326.

Gorry, M. C., Gabbaizedeh, D., Furey, W., Gates, L. K., Preston, R. A., Ashton, C. E., et al. (1997). Mutations in the cationic trypsinogen gene are associated with recurrent acute and chronic pancreatitis. *Gastroenterology, 113*, 1063–1068.

Goto, K., Imura, T., & Tomonaga, M. (2012). Perception of emergent configurations in humans (*Homo sapiens*) and chimpanzees (*Pan troglodytes*). *Journal of Experimental Psychology: Animal Behavioral Processes, 38*, 125–138.

Green, R. E., Krause, J., Briggs, A. W., Maricic, T., Stenzel, U., Kircher, M., et al. (2010). A draft sequence of the Neandertal genome. *Science, 328*, 710–722.

Hattori, Y., Kano, F., & Tomonaga, M. (2010). Differential sensitivity to conspecific and allospecific social cues in chimpanzees (*Pan troglodytes*) and humans (*Homo sapiens*): a comparative eye-tracking study. *Biology Letters, 6*, 610–613.

Hauser, M. D., Chomsky, N., & Fitch, T. (2002). The faculty of language: what is it, who has it, and how did it evolve? *Science, 298*, 1569–1579.

Hayes, C. (1951). *The ape in our house*. New York: Harper.

Hey, J. (2010). The divergence of chimpanzee species and subspecies as revealed in multi-population isolation-with-migration analyses. *Molecular and Biological Evolution, 27,* 921–933.

Hilliard, M. A., Bergamasco, C., Arbucci, S., Plasterk, R. H., & Bazzicalupo, P. (2004). Worms taste bitter: ASH neurons, QUI-1, GPA-3 and ODR-3 mediate quinine avoidance in *Caenorhabditis elegans*. *EMBO Journal, 23,* 1101–1111.

Hong, K. W., Weiss, A., Morimura, N., Udono, T., Hayasaka, I., Humle, T., et al. (2011). Polymorphism of the tryptophan hydroxylase 2 (TPH2) gene is associated with chimpanzee neuroticism. *PLoS One, 6,* e22144.

Imai, H. (in press). In Evolution and Senses: Opsins, Bitter Taste, Olfaction. Springer.

Imai, H., Suzuki, N., Ishimaru, Y., Sakurai, T., Yin, L., Pan, W., et al. (2012). Functional diversity of bitter taste receptor TAS2R16 in primates. *Biology Letters, 8,* 652–656.

Inoue-Murayama, M., Inoue, E., Watanabe, K., Takenaka, A., & Murayama, Y. (2010). Behavior-related candidate genes in Japanese macaques. In N. Nakagawa, M. Nakamichi, & H. Sugiura (Eds.), *The Japanese macaques* (pp. 293–301). Tokyo: Springer.

Inoue-Murayama, M., Mishima, N., Hayasaka, I., Ito, S., & Murayama, Y. (2006). Divergence of ape and human monoamine oxidase A gene promoters: comparative analysis of polymorphisms, tandem repeat structures and transcriptional activities on reporter gene expression. *Neuroscience Letters, 405,* 207–211.

International Human Genome Sequencing Consortium. (2001). Initial sequencing and analysis of the human genome. *Nature, 409,* 860–921.

International Human Genome Sequencing Consortium. (2004). Finishing the euchromatic sequence of the human genome. *Nature, 431,* 931–945.

Kaneko, T., & Tomonaga, M. (2011). The perception of self-agency in chimpanzees (*Pan troglodytes*). *Proceedings of the Royal Society of London. Series B, 278,* 3694–3702.

Kano, F., Hirata, S., Call, J., & Tomonaga, M. (2011). The visual strategy specific to humans among hominids: a study using the gap-overlap paradigm. *Vision Research, 51,* 2348–2355.

Kano, F., & Tomonaga, M. (2009). How chimpanzees look at pictures: a comparative eye-tracking study. *Proceedings of the Royal Society of London. Series B, 276,* 1949–1955.

Kano, F., & Tomonaga, M. (2011). Species difference in the timing of gaze movement between chimpanzees and humans. *Animal Cognition, 14,* 879–892.

Kawai, N., Morokuma, S., Tomonaga, M., Horimoto, N., & Tanaka, M. (2004). Associative learning and memory in a chimpanzee fetus: learning and long lasting memory before birth. *Developmental Psychobiology, 44,* 116–122.

Kellogg, W. N., & Kellogg, L. A. (1933). *The ape and the child: a study of environmental influence upon early behavior*. Oxford, UK: Whittlesey House. 341 pp.

Kobayashi, H., & Kohshima, S. (1997). Unique morphology of the human eye. *Nature, 387,* 767–768.

Konopka, G., & Geschwind, D. H. (2010). Human brain evolution: harnessing the genomics (r) evolution to link genes, cognition, and behavior. *Neuron, 68,* 231–244.

Levy, S., Sutton, G., Ng, P. C., Feuk, L., Halpern, A. L., Walenz, B. P., et al. (2007). The diploid genome sequence of an individual human. *PLoS Biology, 5,* e254.

Lindblad-Toh, K., Wade, C. M., Mikkelsen, T. S., Karlsson, E. K., Jaffe, D. B., Kamal, M., et al. (2005). Genome sequence, comparative analysis and haplotype structure of the domestic dog. *Nature, 438,* 803–819.

Matsui, A., Go, Y., & Niimura, Y. (2010). Degeneration of olfactory receptor gene repertories in primates: no direct link to full trichromatic vision. *Molecular and Biological Evolution, 27,* 1192–1200.

Matsunami, H., Montmayeur, J. P., & Buck, L. B. (2000). A family of candidate taste receptors in human and mouse. *Nature, 404,* 601–604.

Matsuno, T., & Tomonaga, M. (2006). Visual search for moving and stationary items in chimpanzees (*Pan troglodytes*) and humans (*Homo sapiens*). *Behavioural Brain Research, 172,* 219–232.

Matsuzawa, T. (1985). Use of numbers by a chimpanzee. *Nature, 315,* 57–59.

Matsuzawa, T. (1991). *The visual world of chimpanzees*. Tokyo: University of Tokyo Press (Japanese text).

Matsuzawa, T. (2003). The Ai project: historical and ecological contexts. *Animal Cognition, 6*, 199–211.

Matsuzawa, T. (2006). Evolutionary origins of the human mother–infant relationship. In T. Mastuzawa, M. Tomonaga, & M. Tanaka (Eds.), *Cognitive development in chimpanzees* (pp. 127–141). Tokyo: Springer.

Matsuzawa, T., Tomonaga, M., & Tanaka, M. (Eds.). (2006). *Cognitive development in chimpanzees*. Tokyo: Springer.

Meyerhof, W., Batram, C., Kuhn, C., Brockhoff, A., Chudoba, E., Bufe, B., et al. (2010). The molecular receptive ranges of human TAS2R bitter taste receptors. *Chemical Senses, 35*, 157–170.

Mikkelsen, T. S., Batram, C., Kuhn, C., Brockhoff, A., Chudoba, E., Bufe, et al. (2007). Genome of the marsupial *Monodelphis domestica* reveals innovation in non-coding sequences. *Nature, 447*, 167–177.

Miles, H. L. (1999). Symbolic communication with and by great apes. In S. T. Parker & R. W. Mitchell (Eds.), *The mentalities of gorillas and orangutans: Comparative perspectives* (pp. 197–210). NY: Cambridge University Press.

Mizuno, Y., Takeshita, H., & Matsuzawa, T. (2006). Behavior of infant chimpanzees during the night in the first 4 months of life: smiling and suckling in relation to behavioral state. *Infancy, 9*, 215–234.

Mouse Genome Sequencing Consortium. (2002). Initial sequencing and comparative analysis of the mouse genome. *Nature, 420*, 520–562.

Mueller, K. L., Hoon, M. A., Erlenbach, I., Chandrashekar, J., Zuker, C. S., & Ryba, N. J. (2005). The receptors and coding logic for bitter taste. *Nature, 434*, 225–229.

Murofushi, K., Matsuzawa, T., & Asano, T. (1988). Acquisition of agent-action-recipient word order in the chimpanzee. *Primate Research, 4*, 181.

Myowa-Yamakoshi, M., Yamaguchi, M., Tomonaga, M., Tanaka, M., & Matsuzawa, T. (2005). Development of face recognition in infant chimpanzees (*Pan troglodytes*). *Cognitive Development, 20*, 49–63.

Niimura, H. (2012). Evolution of chemosensory receptor genes in primates and other mammals. In H. Hiai, H. Imai, & Y. Go (Eds.), *Post-genome biology of primates* (pp. 43–62). Tokyo: Springer.

Okamoto, S., Tomonaga, M., Ishii, K., Kawai, N., Tanaka, M., & Matsuzawa, T. (2002). An infant chimpanzee (*Pan troglodytes*) follows human gaze. *Animal Cognition, 5*, 97–114.

Okamoto-Barth, S., Kawai, N., Tanaka, M., & Tomonaga, M. (2007). Looking complements distance between mother and infant chimpanzee. *Developmental Science, 10*, 172–182.

Parry, C. M., Erkner, A., & le Coutre, J. (2004). Divergence of T2R chemosensory receptor families in humans, bonobos, and chimpanzees. *Proceedings of the National Academy of Sciences United States of America, 101*, 14830–14834.

Patterson, F. G., & Linden, E. (1981). *The education of Koko*. New York: Holt, Rinehart and Winston.

Premack, D. (1976). *Intelligence in ape and man*. Hillsdale: Lawrence Erlbaum.

Rat Genome Sequencing Project Consortium. (2004). Genome sequence of the Brown Norway rat yields insights into mammalian evolution. *Nature, 428*, 493–521.

Rhesus Macaque Genome Sequencing and Analysis Consortium. (2007). Evolutionary and biomedical insights from the rhesus macaque genome. *Science, 316*, 222–234.

Rumbaugh, D. M. (Ed.). (1977). *Language learning by a chimpanzee: The LANA Project*. New York: Academic Press.

Savage-Rumbaugh, E. S., & Lewin, R. (1994). *The ape at the brink of the human mind*. Kanzi: John Wiley and Sons.

Shi, P., & Zhang, J. (2006). Contrasting modes of evolution between vertebrate sweet/umami receptor genes and bitter receptor genes. *Molecular and Biological Evolution, 23*, 292–300.

Shi, P., Zhang, J., Yang, H., & Zhang, Y. P. (2003). Adaptive diversification of bitter taste receptor genes in mammalian evolution. *Molecular and Biological Evolution, 20*, 805–814.

Shimada, M. K., Inoue-Murayama, M., Ueda, Y., Maejima, M., Murayama, Y., Takenaka, O., et al. (2004). Polymorphism in the second intron of dopamine receptor D4 gene in humans and apes. *Biochemical and Biophysical Research Communications, 316*, 1186–1190.

Sidman, M., Tailby, W., Sidman, M., & Tailby, W. (1982). Conditional discrimination vs. matching to sample: an expansion of the testing paradigm. *Journal of Experimental Analysis of Behavior, 37*, 5–22.

Sugawara, T., Go, Y., Udono, T., Morimura, N., Tomonaga, M., Hirai, H., et al. (2011). Diversification of bitter taste receptor gene family in western chimpanzees. *Molecular and Biological Evolution, 28*, 921–931.

Sugawara, T., & Imai, H. (2012). Post-genome biology of primates focusing on taste perception. In H. Hirai, H. Imai, & Y. Go (Eds.), *Post-genome biology of primates* (pp. 79–92). Tokyo: Springer.

Suzuki, N., Sugawara, T., Matsui, A., Go, Y., Hirai, H., & Imai, H. (2010). Identification of non-taster Japanese macaques for a specific bitter taste. *Primates, 51*, 285–289.

Terrace, H. S. (1979). *Nim: A chimpanzee who learned sign language*. New York: Alfred A. Knopf.

Terrace, H. S. (2011). Missing links in the evolution of language. In S. Dehaene & Y. Christen (Eds.), *Characterizing consciousness: From cognition to the clinic?* (pp. 1–25). Heidelberg: Springer.

The 1000 Genomes Project Consortium. (2010). A map of human genome variation from popula-tion-scale sequencing. *Nature, 467*, 1061–1073.

Tomasello, M. (1999). *The cultural origins of human cognition*. Cambridge: Harvard University Press.

Tomonaga, M. (1997). Precuing the target location in visual searching by a chimpanzee *Pan trog-lodytes*: Effects of precue validity. *Japanese Psychological Research, 39*, 200–211.

Tomonaga, M. (1999a). Inversion effect in perception of human faces in a chimpanzee (*Pan trog-lodytes*). *Primates, 40*, 417–438.

Tomonaga, M. (1999b). Visual search for orientation of faces by a chimpanzee (*Pan troglodytes*). *Primate Research, 15*, 215–229.

Tomonaga, M. (2001a). Investigating visual perception and cognition in chimpanzees (*Pan troglo-dytes*) through visual search and related tasks: From basic to complex processes. In T. Matsuzawa (Ed.), *Primate origins of human cognition and behavior* (pp. 55–86). Tokyo: Springer.

Tomonaga, M. (2001b). Visual search for biological motion patterns in chimpanzees (*Pan troglo-dytes*). *Psychologia, 44*, 46–59.

Tomonaga, M. (2007a). Visual search for orientation of faces by a chimpanzee (Pan troglodytes): face-specific upright superiority and the role of configural properties of faces. *Primates, 48*, 1–12.

Tomonaga, M. (2007b). Is chimpanzee (*Pan troglodytes*) spatial attention reflexively triggered by the gaze cue? *Journal of Comparative Psychology, 121*, 156–170.

Tomonaga, M. (2008). (Non-) emergence of symmetry in chimpanzees. *Cognitive Studies, 15*, 347–357 (Japanese text with English summary).

Tomonaga, M. (2010). Do the chimpanzee eyes have it? In E. V. Lonsdorf, S. R. Ross, & T. Matsuzawa (Eds.), *The mind of the chimpanzee: Ecological and empirical perspectives* (pp. 42–59). Chicago: The University of Chicago Press.

Tomonaga, M., & Imura, T. (2009a). Human gestures trigger different attentional shifts in chim-panzees (Pan *troglodytes*) and humans (*Homo sapiens*). *Animal Cognition, 12*, S11–S18.

Tomonaga, M., & Imura, T. (2009b). Faces capture the visuospatial attention of chimpanzees (*Pan troglodytes*): evidence from a cueing experiment. *Frontiers in Zoology, 6*, 14. doi:10.1186/1742-9994-6-14.

Tomonaga, M., Myowa-Yamakoshi, M., Mizuno, Y., Okamoto, S., Yamaguchi, M. K., Kosugi, D., et al. (2004). Development of social cognition in infant chimpanzees (*Pan troglodytes*): face recognition, smiling, gaze and the lack of triadic interactions. *Japanese Psychological Research, 46*, 227–235.

Tomonaga, M., Tanaka, M., & Matsuzawa, T. (Eds.). (2003). *Cognitive and behavioral develop-ment in chimpanzees*. Kyoto: Kyoto University Press.

Ueno, A., Ueno, Y., & Tomonaga, M. (2004). Facial responses to four basic tastes in newborn rhesus macaques (*Macaca mulatta*) and chimpanzees (*Pan troglodytes*). *Behavioral and Brain Research, 154*, 261–271.

Venter, J. C., Adams, M. D., Myers, E. W., Li, P. W., Mural, R. J., Sutton, G. G., et al. (2001). The sequence of the human genome. *Science, 291*, 1304–1351.

Wang, J., Wang, W., Li, R., Li, Y., Tian, G., Goodman, L., et al. (2008). The diploid genome sequence of an Asian individual. *Nature, 456*, 60–65.

Wheeler, D. A., Srinivasan, M., Egholm, M., Shen, Y., Chen, L., McGuire, A., et al. (2008). The complete genome of an individual by massively parallel DNA sequencing. *Nature, 452*, 872–876.

Whitcomb, D. C., Gorry, M. C., Preston, R. A., Furey, W., Sossenheimer, M. J., Ulrich, C. D., et al. (1996). Hereditary pancreatitis is caused by a mutation in the cationic trypsinogen gene. *Nature Genetics, 14*, 141–145.

Wooding, S., Bufe, B., Grassi, C., Howard, M. T., Stone, A. C., Vazquez, M., et al. (2006). Independent evolution of bitter-taste sensitivity in humans and chimpanzees. *Nature, 440*, 930–934.

Yamamoto, J., & Asano, T. (1995). Stimulus equivalence in a chimpanzee. *Psychological Record, 45*, 3–21.

Yu, N., Jensen-Seaman, M. I., Chemnick, L., Ryder, O., & Li, W. H. (2004). Nucleotide diversity in gorillas. *Genetics, 166*, 1375–1383.